Ferdinand von Hochstetter

The Geology of New Zealand

In explanation of the geographical and topographical atlas of New Zealand, from the scientific publications of the Novara Expedition

Ferdinand von Hochstetter

The Geology of New Zealand
In explanation of the geographical and topographical atlas of New Zealand, from the scientific publications of the Novara Expedition

ISBN/EAN: 9783337321802

Printed in Europe, USA, Canada, Australia, Japan

Cover: Foto ©Andreas Hilbeck / pixelio.de

More available books at **www.hansebooks.com**

THE

GEOLOGY OF NEW ZEALAND:

IN EXPLANATION OF THE

GEOGRAPHICAL AND TOPOGRAPHICAL

ATLAS OF NEW ZEALAND

BY

DR. F. VON HOCHSTETTER AND DR. A. PETERMANN.

FROM THE SCIENTIFIC PUBLICATIONS OF THE
NOVARA EXPEDITION.

TRANSLATED BY DR. C. F. FISCHER.

ALSO,

LECTURES BY DR. F. HOCHSTETTER

DELIVERED IN NEW ZEALAND.

AUCKLAND: T. DELATTRE, QUEEN STREET.

1864.

CONTENTS.

	PAGE.
Observations upon the Chartography of New Zealand, by Dr. A. Petermann	3
Lecture on the Geology of the Province of Auckland, by Dr. F. v. Hochstetter	8
New Zealand, Geographical and Geological Survey: Explanation of Map I.	43
The Geological Formation of the Southern Part of the Province of Auckland, Map II.	47
Record of the heights of the Southern part of the Province of Auckland	56
The Isthmus of Auckland and its extinct Volcanos, Map III.	62
Rotomahana (or the Warm Lake) and its hot springs, Map IV.	67
Whaingaroa, Aotea, and Kawhia, three harbours on the West Coast of the Province of Auckland, Map V.	72
Lecture on the Geology of the Province of Nelson, by Dr. F. v. Hochstetter	77
Explanation of the Map of the Province of Nelson, Map VI.	108

I.
OBSERVATIONS UPON THE CHARTOGRAPHY OF NEW ZEALAND.

BY DR. A. PETERMANN.

THE History and Progress of the Geographical Knowledge and Chartography of New Zealand may be classified into four periods :—
    1642, The discovery by Tasman.
    1769, The investigation and survey by Cook.
    1848, Survey by the English Admiralty.
    1859, Commencement of the surveys in the interior by F. von Hochstetter and Julius Haast.

The Dutch navigator, Abel Jansen Tasman, discovered New Zealand on the 13th December, 1642, observing from the westward the clouded summits of the Southern Alps. He sailed along the coast, passing Cook's Straits and the Northern Island up to the Three Kings. Although he saw the greater part of the West Coast of New Zealand, the result of his observations was very incomplete and erroneous, which is proved by the fact that he considered New Zealand as a part of the *Terra Australis Incognita* which, according to his supposition, stretched to the far east, and was connected with the South Cape of America.

The knowledge of New Zealand made no advance for nearly a century, until the time when Cook anchored at Tauranga, Poverty Bay, on the East Coast, on the 8th of October, 1769; and it was on this his first visit and his second and third (1773-74, 1779), that he investigated New Zealand, sailed round it, and finished a survey of its entire coast. New Zealand was visited nearly at the same time by two French navigators, viz. :—in December, 1769, by Captain Durville, and in the year 1772 by the unfortunate Captain Marion, who was killed and eaten by the natives at the Bay of Islands. Neither of the expeditions added anything of importance to Cook's observations or to the knowledge of the country.

Through Cook's glorious discoveries the attention of Europe was drawn in a very marked manner to New Zealand. Whalers visited its harbours, and occasional adventurers began to settle; but the early period of the European colonization was attended only with crime and disgrace. A new and better era began with the year 1814, when Samuel Marsden founded the first Christian mission; from that time the intercourse between the Europeans and Aboriginals was better regulated. An attempt at colonization was made in the year 1825, but it was not until 1840 that New Zealand became an English Colony.

Since the time of Cook, in the year 1769, and still more, since that of Marsden, in 1814, down to the Admiralty surveys in 1848, the literature bearing upon New Zealand is comprised in a great number of very valuable publications, official reports, works of travels, books, pamphlets of various kinds, charts and maps. Thompson* counts not less than two hundred and forty-five. Amongst the maps of the period are Cook's surveys, the detail charts of separate bays and harbours, by English and French naval officers—reckoning from North to South: Port Monganui, by A. H. Halloran, 1845; Bay of Islands, by M. Duperrey, 1824; Tutukaka Harbour and Nongodo River, by N. C. Phillips, 1837; Mahurangi Harbour, by J. A. Cudlip, 1834; Port Nicholson, by E. M. Chaffers, 1839; Manukau Harbour, by G. O. Ormsby, 1845; Torrent Bay and Astrolabe Road, by M. Guilbert, 1827; Current Basin, by M. Guilbert, 1827; Port Hardy and Port Gore, by Lieutenant Moore, 1834; Tory Channel, by E. M. Chaffers, 1839; Port Underwood in Cloudy Bay, by G. Johnson, 1837; Akaroa Harbour, by Commander O. Stanley, 1840; Rouabouki Road, by Lieut. O. Wilson, 1839; Dusky and Chalky Bay, by M. Duperrey, 1824. The survey of the settlement and the beautiful map of New Plymouth and its vicinity, by F. A. Carrington, 1840;

---

* "The Story of New Zealand," vol. II., p. 341 *et seq.*; see also the larger work by Hochstetter—"New Zealand," p. 549.

the Harbour and City of Auckland, the Capital of New Zealand, with the districts of the rivers Kaipara, Waitemata, Tamaki, Wairoa, Waihou or Thames, Mercury Bay, Kawia, Piako, Waipa, Waikato, Manukau, Tauranga, etc., compiled from various sources by J. Arrowsmith, 1842 (with branch maps: Auckland the Capital of New Zealand, surveyed by Felton Matthew, Surveyor-General of New Zealand, 1841; and a Trigonometrical Survey of the Harbour of the Waitemata and the Isthmus which separates the waters of the Thames from those of the Manukau, by Captain Owen Stanley, R.N., and Felton Matthew, 1841); the maps of Dieffenbach's travels, by Arrowsmith, &c.

All the maps and surveys of New Zealand existing before 1848, consisted of disconnected fragments; but it must be remembered that it is a country of an area of not less than 630 square miles larger than the present Kingdom of Italy.*

The important survey of the New Zealand coast was undertaken by the command of the English Admiralty, under the direction of Captain J. Lort Stokes and Commander Byron Drury, in the surveying ships "Acheron" and "Pandora," and occupied a period of eight years, from 1848 to 1855, and now forms one of the most magnificent of the many productions of the English navy. Besides the above-named commanders, the following officers were engaged in this work: Commander G. H. Richards, F. J. Evans, R. Bradshaw, J. W. Smith, P. W. Oke, R. Burnett, H. Kerr, T. Kerr, W. Blackney, H. Ellis, A. Farmer, C. Stanley, J. M. Pridham, D. Pender, J. W. Hamilton, and C. Kettle. The result of this survey occupies fifty sheets, thirty-three of which are in the large chart formula (double elephant). They are carefully engraved on copper, and were published in twelve years, from 1850 to 1861, in the following series:—fourteen sheets were published from 1850 to 1856, twenty-one sheets in the year 1857, and twelve from 1858 to 1861. Some sheets contain several plans in various scales, from the smallest 1:1,750,000 to the largest 1:5000. In the smallest size is published the general chart,

---

* The area of New Zealand is 4,703 German, and that of Italy 4,674 German square miles.

No. 1,212—price, 3s. 6d. Then follow fourteen sheets of the largest formula, which are on one and the same scale of 1:280.000,* and which embrace the whole of New Zealand.† Comparing these fourteen sheets with Reymann's Map of Germany and Central Europe, on the scale of 1:200.000, and supposing them divided in the same manner as Reymann's, they would have formed exactly one hundred of such sheets. This will give an idea of the magnitude of this survey. Of the remaining fifty-nine charts and plans, six are at a scale of 1:145.000 to 1:48.000; eleven, 1:36.000; two, 1:27.000 and 1:26.000; seventeen, 1:24.000; three, 1:22.000; eleven, 1:21.000 to 1:12.000; and nine, 1:9000 to 1:5000.

While, through the surveys of the Admiralty, the outlines of New Zealand were carefully and completely settled, the knowledge of the interior was gradually developed by the surveys of the various settlements and through exploring expeditions, and especially in the South Island journeys of discovery were undertaken into the interior. The account of the expedition of Thomas Brunner, in the year 1846-47, from Nelson in a southerly direction along the coast to Tihihai Head, was published in the Journal of the Royal Geographical Society, in 1850, with a small map. A short account of

---

* As all the charts are according to Mercator's projection, the single sheets differ naturally in the scales, and in such a way that the northern sections are 1:300.000, and the southern 1:260.000.

† The sections have the following number and outlines:—
Sheet 1, No. 2525, The northern coasts from Hokianga on the west to Tukukaka on the east.
Sheet 2, No. 2543, Monganui Bluff to Manukau on the West Coast, and from Tukukaka to Major Island on the East Coast.
Sheet 3, No. 2527, Major Island to Poverty Bay.
Sheet 4, No. 2521, Poverty Bay to Cape Palliser.
Sheet 5, No. 2054, Cook's Strait and Coast to Cape Egmont.
Sheet 6, No. 2535, Manukau Harbour to Cape Egmont.
Sheet 7, No. 2616, Cape Foulwind to D'Urville Island, including Blind and Massacre Bays.
Sheet 8, No. 2529, Cape Campbell to Banks' Peninsula.
Sheet 9, No. 2532, Ninety Miles Beach to Otago.
Sheet 10, No. 2533, Otago to Mataura River and Ruapuke Island.
Sheet 11, No. 2553, Foveaux Strait and Stewart's Island.
Sheet 12, No. 2589, Foveaux Strait to Awarua River.
Sheet 13, No. 2590, Awarua River to Abut Head.
Sheet 14, No. 2591, Abut Head to Cape Foulwind.
(The price of each sheet is 2s. 6d.)

Dashwood and Mitchell's tour from Nelson to Lyttelton, along with a small map, appeared in the volume for 1851. J. T. Thompson's account and map of what is now the Province of Southland in the volume for 1858. E. Stanford published in London, in the year 1856, a map of the Province of Canterbury, showing freehold sections and pasturage runs; scale, 1:220.000. J. Arrowsmith published two editions of his map of New Zealand on a scale of 1:2.400.000—one dated July, 1851, the other, July, 1858.

With the explorations and surveys of F. v. Hochstetter and J. Haast (which were commenced in 1859, and are still unfinished), began a new epoch in the geographical knowledge and chartography of New Zealand.* Not only have their labours enlarged the existing knowledge, but they have thrown quite a new light on the geological and topographical condition of the interior, as hitherto the topographical configuration of the country has been much neglected by the colonial surveys. The general map of this work (Map I.), in the completion of which Hochstetter's and Haast's observations have been used for the first time, will show at the first glance how much our former conception of New Zealand is enlarged and corrected. Of course, on it the observations of many others also have been recorded.

The progress of the chartography of New Zealand is best shown in the various editions of Arrowsmith's maps, of which there are three—1841, 1851, and 1858—which were compiled from official and other documents existing at those times. The edition of 1841 contains nothing but a coast line, and this very imperfect—here and there an error of half a degree; in the interior are only a few roughly noted lakes, rivers, and mountains. In the edition of 1851 the coast, if yet incomplete, is corrected after the marine surveys, and the interior is filled up. But the edition of 1858 contains many additions

---

* A report by Dr. J. Haast, Geologist of the Province of Canterbury, dated 3rd of March, 1863, describes his latest travels and surveys of the Southern Alps. This traveller had penetrated into the upper region of the Molyneaux River, with its magnificent lakes Wanaka and Hawea, to the West Coast, and discovered a pass through the chain of Alps at an elevation of only 1,612 feet, between the Wanaka Lake and the Awarua River.

of importance. The most complete map published before ours appeared in the Despatch Atlas, March, 1861, in two sheets, and at a scale of 1:1.900.000 and 1:2.300.000.

According to our map the area of New Zealand may be estimated according to planimetric calculation, at—

North Island... 2041·6 German, or 43,400·9 Eng. sq. miles.
South Island... 2627·7   ,,   55,860·3   ,,
Stewart's Island 33·3   ,,   707·8   ,,

             4702·6   99,969*

II.
LECTURE ON THE GEOLOGY OF THE PROVINCE OF AUCKLAND.

BY DR. F. VON HOCHSTETTER.

[Delivered to the Members of the Auckland Mechanics' Institute, June 24, 1859.]

MR. PRESIDENT, LADIES AND GENTLEMEN,—

The members of the Auckland Mechanics' Institute having done me the honour to elect me as honorary member of their institution, and the Committee having invited me to give a Lecture upon the Geology of this Province, I have much pleasure in complying with their request. It is, however, with some

---

* From English sources the area and circumference of New Zealand has been estimated at—

	SURFACE.		CIRCUMFERENCE OF COAST.
	Acres	Sq. Mls.	Sea Miles
North Island........	31,174,400	48,710	1,500
Middle Island	46,126,860	72,072	1,500
Stewart's Island ...	1,152,000	1,800	120
	78,452,480	122,582	3,120

The area of New Zealand is 50,000 acres less than the area of Great Britain and Ireland; the Northern Island is 1-32nd smaller than England; the Middle Island is 1-9th smaller than England and Scotland combined; 50,000,000 acres, or 2-3rds, of which are estimated fit for agriculture; the rest is made up of impassable mountains, sand-flats, swamps, lakes, and rivers.—(*Vide* "New Zealand," by J. v. Hochstetter.)

hesitation that I undertake this task, feeling that my imperfect knowledge of the English language will prevent my making the short sketch I wish to lay before you as interesting as it might otherwise have been. Notwithstanding this drawback, I am glad to have this opportunity of giving the inhabitants of this Province, through the members of this Institute, such a *résumé* as I can of the chief results of the Geological Survey I have made of those parts of the country I have visited.

I feel this, indeed, to be a duty I owe to the community at large, in return for the very kind reception that has everywhere been given me—for the ready help that has always been afforded by all whom I have met with—and for the interest that has been shown by all in the proceedings of the Imperial Austrian "Novara" Expedition.

Having, in the months of January and February, completed my Survey, and finished a Geological Map, of the Auckland District—which I now have the pleasure of showing you—the necessity arose for my choosing either the *Northern* or the *Southern* portion of the Province for my further researches, my limited stay in New Zealand rendering it impossible for me to make a sufficient examination in both directions.

I did not hesitate to choose the Southern districts—for these reasons: that the country over which I should there proceed, is inhabited almost exclusively by Maoris, and has hitherto been almost unknown and totally unsurveyed, both topographically and geologically. The Northern districts, on the contrary, are for the most part better known, and from the number of European settlers in them, I was led to hope I should be enabled to collect some information through specimens forwarded to me for examination, and from the verbal descriptions of those who are well acquainted with the various localities.

My hope was not unfounded in either respect.

I have received many specimens of interest from various localities; also some valuable information from different settlers, and especially from my friends, the Rev. A. G. Purchas and Mr. C. Heaphy, who in the last few months have had opportunities of visiting several parts of the Northern portion of this Province, and of collecting very valuable specimens. In addition to this must be remembered the fact, that other scientific men, especially MM. Dieffenbach and Dana, had already visited and decribed at length some parts of the Northern country.

Through the liberality and excellent arrangements of the General and Provincial Governments, I have been enabled in a comparatively short time to travel over and to examine the larger portion of the Province South of Auckland, extending as far as Lake Taupo and Tongariro Volcano, the boundaries between this Province and those of Wellington and Hawke's Bay. I have thus obtained materials which will enable me, on my return to Europe, to construct a Topographical and Geological Map of the central part of the Northern Island.

My observations have, with the able assistance of Mr. Drummond Hay, extended from the East to the West Coast; and the numerous peaks and ranges have afforded facilities for fixing with satisfactory accuracy, by means of *magnetic bearings*, on the basis of points previously fixed by the nautical survey of Capt. Drury on the coast-line, all the great natural features of this portion of the country. A great number of *barometrical observations* have afforded me the means of ascertaining the heights of mountains and plains in the interior, which I shall be able to calculate with accuracy by the aid of corresponding daily observations, taken in Auckland by Colonel Mould, who has kindly forwarded me a copy of his tables.

I have also obtained *photographic* and other *views* of great interest, many of which were taken by the gentlemen who accompanied me on the expedition for this purpose; and a large number of exceedingly valuable sketches have been contributed by the talented pencil of our president, Mr. C. Heaphy, for future publication in a geological atlas. Many of these are decorating the walls and others are lying on the table, and I shall be happy to show them to any ladies and gentlemen who may feel any interest in seeing them, at the conclusion of the lecture.

My *collections* have been growing from day to day, and include specimens of great interest in most branches of Natural History. I owe a great deal to the indefatigable zeal of my friend and fellow-traveller, Mr. J. Haast, who assisted me in collecting during our expedition. I am also much indebted to Mr. J. Crawford at Wellington, Mr. A. S. Atkinson of Taranaki, Mr. Triphook of Hawke's Bay, Mr. H. T. Kemp of the

Bay of Islands, to the Missionaries, and to almost innumerable friends in Auckland.

PRELIMINARY REMARKS.

I cannot suppose that all my audience are acquainted with the first principles of Geology. I shall therefore be under the necessity, in order to make my report intelligible, of prefacing a few remarks upon the chief divisions of the geological formations.

The various rocks, soils, and minerals which occur upon the surface of the earth, or at various depths beneath it—in one word, the materials of the "*earth's crust*"—are classified, in the first place, with reference to their different *origin*, or, in other words, with reference to the different circumstances and causes by which they have been produced. They are divided into *four* great classes—*Plutonic, Metamorphic, Aqueous,* and *Volcanic* rocks. Another mode of classification is with reference to their *age*—that is, to the comparative periods of their formation. Those divisions will be easily understood.

The *Plutonic* rocks comprehend all the *granites, syenites, porphyries, diorites*—rocks which agree in being highly crystalline, unstratified, and destitute of organic remains—which are considered as of igneous origin, formed in the earliest periods of the earth, in great depths, and cooled and crystalised slowly under great pressure.

The *Metamorphic* rocks are the crystalline strata, or schists, called *gneiss, mica-schist,* or *mica-slate, chlorite-schist, hornblende-schist*—also destitute of organic remains. According to the most probable theory, these strata were originally deposited from water in the usual form of sediment, but were subsequently altered by subterranean heat, so as to assume a new texture.

The two first classes of rocks are usually found in such a position that they form the foundation on which the aqueous rocks were afterwards superimposed. For instance, they compose the central line or a range of mountains, on both sides of which sedimentary rocks are deposited. Thus, in reference to their *age*, they are considered as the oldest, and are therefore called also *Primitive*.

There are exceptions to this rule in reference to the age of certain plutonic rocks of *eruptive* character. But I am now stating only general principles, and therefore avoid all questions leading to scientific discussions.

The next in order are the *Aqueous* rocks—the production of watery action. They are also called *sedimentary* rocks, from the fact that they are the hardened sediments accumulated at the bottom of the sea or of fresh-water lakes. They are stratified, or divided into distinct layers of strata : as, for example, clay-slate, marl, sandstone, limestone, and are divided into three kinds, called *arenaceous* or *siliceous*, *argillaceous* or *clayey*, and *calcareous* or *chalky*—according to the respective predominance of silica, alumina, or lime. Rocks of this class cover a larger part of the earth's surface than any others, and are of the greatest interest on account of the *organic* remains which are found imbedded in the different strata.

There are two principal means of ascertaining the relative age of aqueous rocks—derived, the one from their *position*, the other from the *fossil remains* they contain.

With reference to *position*—the bed which lies uppermost is of course the newest of all, and that which lies at the bottom, the most ancient.

With reference to the *fossils*, it is not so easy to give an explanation in few words; but some idea may be formed from the well-ascertained fact, that certain animals have existed for a certain period, and then wholly disappeared and been succeeded by other animals of a different species, which, in turn, have again given place to others.

So, as Sir Charles Lyell truly says, "a series of sedimentary formations is like volumes of history, in which each writer has recorded the annals of his own times, and then laid down the book with the last written page uppermost." And the organic remains are, as Dr. Mantell beautifully expresses it, the "coins of Creation," which give us the means of tracing the history of the development of the organic kingdoms.

Thus, by superposition and by their organic remains, the aqueous rocks are divided into groups forming, in reference to their age, what is termed an "ascending series," or beginning with the oldest, in the following manner :—

1. Primary formations or periods
2. Secondary ,, ,,
3. Tertiary ,, ,,
4. Quartary ,, ,,

In reference to the word "quartary" I may explain that, although it is not an English word, I take the liberty to use it in the sense of "post tertiary," as following the analogy of the other terms.

Each of those formations is again divided into numerous minor systems, on which I have no time to enter.

The fourth and last great divisions of rocks are the *volcanic* —as Trachyte, Basalt, Breccia, and Tuff—all produced by supramarine or submarine volcanic eruption. It is ascertained that the earliest true volcanic eruptions have occurred subsequently to the Secondary period, commencing in the Tertiary, and continuing to the present time; and there is a marked difference between the older and the more recent eruptions.

I have prepared a diagram which will serve to impress these first principles upon your memory, and so enable you to follow me in the account I have to give :—

DIAGRAM.

Origin.	Age.	Organic Remains.
Plutonic and Metamorphic rocks	Primitive formation.	No fossils.
Aqueous	Primary. Secondary. Tertiary. Quartary.	Fossilferous.
Volcanic	Trachytic. Basaltic.	No fossils.

With these preliminary remarks, I now proceed to the main subject of my lecture.

GEOLOGY OF THE PROVINCE OF AUCKLAND.

The first striking characteristic of the Geology of this Province—and probably of the whole of the Northern Island of New Zealand—is *the absence of the primitive plutonic and metamorphic* formations, as granite, gneiss, mica-slate, and the like. I have been informed by Mr. Heaphy that these rocks are of wide-spread extent in the Middle Island, forming mountain ranges of great altitude, covered with perpetual snow, and reaching in Mount Cook probably to 13,000 feet. The rocks of these formations contain the principal metallic riches of the earth. Therefore we cannot hope to find these riches developed in the highest degree in the Northern Island; but as other formations also contain metalliferous veins, there may be found many mines worth working in the rocks I am about to describe.

I.—PRIMARY FORMATION.

The oldest rock I have met with in the Province of Auckland belongs to the *primary formation*. It is of very variable character—sometimes being more argillaceous, of a dark blue colour (when decomposed, yellowish brown, the colour generally presented on the surface,) and more or less distinctly stratified like *clay slate*—(at Maraitai on the Waitemata); at other times the siliceous element preponderates, and, from the admixture of oxide of iron, the rock has a red, jasper-like appearance—(at Waiheki, Manganese Point.) In other localities it is more distinctly arenaceous, resembling the Old Sandstones of the Silurian and Devonian Systems, called Grauwacke—(at Taupo, on the Hauraki Gulf.)

As no fossils have yet been found in this formation in New Zealand, it is impossible to state the exact age; I am, however, of opinion that these argillaceous siliceous rocks will be found to correspond to the oldest Silurian strata of Europe.

The existence and great extent of *this* formation are of considerable importance to this Province, as *all the metalliferous veins* hitherto discovered, or likely to be hereafter found, occur in rocks of this formation.

To these rocks belong the *Copper-pyrites*, which has been worked for some years at the *Kawau* and *Great Barrier*—the *Manganese* (Psilomelan) at *Waiheki*—and the *Gold-bearing quartz* at *Coromandel*.

The *gold* which is washed out from beds of quartz-gravel in the rivers and creeks flowing down from both sides of Coromandel range, is derived from quartz veins, of crystalline character and considerable thickness, running, in a general direction from North to South, through the old primary rocks which form the foundation of the Coromandel range. In some places these veins stand up like a wall on the summit of the range to a height of eight or ten feet. The clay-slate rock itself is exposed only at the bottom of deep gorges which form the channels of the principal streams. In almost all places it is covered by large masses of trachytic tuff and breccia, of which the hills surrounding the Harbour of Coromandel are composed. The well-known "Castle Hill"—which can be seen from Auckland—is a characteristic example of the Trachytic Breccia formation. The magnetic iron-sand which, in washing, is found with the gold, is derived from the same source as all the magnetic iron-sand of New Zealand—namely, from the decomposition of trachytic rocks. Small veins of quartz of amorphous character—that is, not crystalline, but in the shape of chalcedony, cornelian, agate, and jasper—are found in numerous places on the shores of Coromandel. These veins occurring in trachytic rocks, are quite different from the auriferous quartz veins in the primary formation—a fact, I think, of much practical importance to state, to prevent the fruitless search for gold where gold does not exist. All the gold-bearing gravel in the creeks is derived, as I have already said, not from the veins in the trachytic breccia, but from the much thicker and crystalline veins in the primary rocks. The surface-deposit in those creeks is very rich, but, as compared with Australian and Californian gold-fields, of limited extent and depth. I washed a few bucketsful of surface earth, and gravel, at a creek pointed out to me by Mr. Charles Heaphy, near Ring's Mill, at the Kapanga. Every panful showed scales of thin gold, with small fragments of quartz streaked and studded with veins and spangles of gold. These "specimens," as they are called by diggers, show no—or very little—sign of being water-worn, but are sharp and crisp fragments, as if they had been broken up on the spot, or in the immediate vicinity. I think the quartz veins in the mountains should be thoroughly examined, and that, when once the day has come that the Coromandel gold-fields are worked, the attention of the "digger" should be directed as well to the hills immediately above any rich deposits as to the alluvial workings below.

The Coal Beds at Coromandel occurring between strata of trachytic breccia are too thin to be of any value, and as the coal formation is absent, there is no ground for hoping that a workable seam may be found.

The primary formation occurs, to a more considerable extent, to the eastward of Auckland, in ranges on both sides of the *Wairoa river* attaining an altitude of 1,500 feet above the sea—and striking from thence northwards, over Waiheki and Kawau, to the Bay of Islands. In a Southerly direction, they extend, through the *Hangawera* and *Taupiri* ranges, across the Waikato, through the *Hakarimata* and *Hauturu* range—parallel with the West Coast—to the Mokau district, where, at Wairere, the Mokau river falls in a magnificent cascade over a lofty precipice of that rock.

The same formation occurs again in the *Rangitoto mountain* on the Upper Waipa, and West of Taupo lake in the *Tuhua* mountains. But the most extensive range of primary rocks is that which commences near Wellington under the name of *Tararua* and *Ruawahine*, and runs in a north-easterly direction to the east shore of Taupo lake, under the name of *Kaimanawa*, in which rises the principal source of the Waikato—there called Tongariro river. The range continues from the shores of Taupo lake, in a north-easterly direction, to the East Cape, under the principal name of *Tewhaiti*. This lofty and extensive mountain range—the true backbone of the Northern Island—with peaks from 6000 to 7000 feet, is entirely unknown. In this range the *Plutonic* and *Metamorphic* rocks, yet unknown in the Northern Island, may perhaps be found.

Nearly all the primary ranges are covered with dense virgin forests, which render them extremely difficult of access. It must be left to the labour and enterprise of future years to discover and develop the mineral riches, the existence of which appears to be probable, not only from the geological characteristics of the country, but also from some few specimens of Lead and Copper ore that have from time to time been picked up by the Natives.

It is remarkable that, while one of the oldest members of the primary formation is found so extensively in New Zealand, the later strata, as the Devonian, Carboniferous, and Permian systems, appear to be altogether wanting;—while, on the other hand, in the neighbouring Continent of Australia these members of the primary period, together with plutonic and metamorphic rocks, constitute, so far as we know, almost the principal part of the continent.

II.—SECONDARY FORMATION.

A very wide interval occurs between the primary rocks of the Northern Island and the next sedimentary strata that I met with. Not only the upper members of the primary series are absent, but also nearly the whole of the Secondary formations. The only instance of secondary strata that I have met with consists of very regular and highly-inclined beds of marl alternating with micaceous sandstone, extending to a thickness of more than 1000 feet—which I first saw on the South head of the Waikato, and afterwards met with on the western shore of Kawhia harbour.

These rocks possess great interest from the fact that they contain remarkable specimens of marine fossils, which belong exclusively to the secondary period, especially Cephalopods of the genera *Ammonite* and *Belemnite*, several species of *Belemnite*, all belonging to the family of the *Canaticulati*. These are the first specimen of those genera which have been discovered in the regions of Australasia. Both fossils have been known for centuries by our ancestors in the Old World—the Ammonite as the horn of Jupiter Ammon, and the Belemnite as the bolts of the God of Thunder. The latter, though now first seen in the Antipodes by Europeans, have long been known to the Natives of Kawhia by a much less dignified name—the old warrior-chief, *Nuitone te Pakaru*, having told me that the stones I prized so much and collected so greedily, are nothing more than '*roke-kanae*,' which means the excrement of the fish commonly known among the settlers by the name of 'mullet.' In reality, the Belemnite belongs to a creature, long since extinct, which was allied to the now living cuttle-fish.

Secondary rocks may probably be found in some other parts of the West Coast, and occur, as I have been kindly informed by the Rev. A. G. Purchas, in the Harbour of Hokianga—but everywhere of limited superficial extent.

III.—TERTIARY FORMATIONS.

I proceed now to speak of the Tertiary period, strata of which, of very various characters, occupy a large portion of the Northern Island. The various tertiary strata are found for the most part in a horizontal position—a remarkable fact, from which we may conclude that even the numerous volcanic eruptions which took place during and after the period of their deposition, had not power enough to dislocate the whole system, but merely to produce local disturbances.

The tertiary period must be divided into two distinct formations, which may perhaps correspond to the European *Eocene* and *Miocene*. There is an older formation which is found principally on the West Coast, and in the interior, on both sides of the primary ranges, and a newer one which may be called the *Auckland Tertiary Formation*.

You will probably be interested to have some more minute description of the different strata of the older of these formations, as to this belong the *Brown-Coal* seams, to the discovery of which I am indebted for the opportunity of investigating the Geology of this Province, and on the intelligent working of which I believe very much of the future welfare of this Province depends.

The *Brown-Coal Formation* is of a very considerable extent, both in the Northern and Middle Islands of New Zealand, and is of similar character everywhere.

Some months ago I furnished a report on the Coalfield in the neighbourhood of Auckland, in the Drury and Hunua districts, of which I will repeat here the principal points. The Drury coal belongs to a very good sort of brown coal—to the so-called *Glanzkohle*, with conchoidal fracture. I was not able to convince myself of the existence of different series of seams, one above the other, on different levels. I am much rather of opinion that the same seam, disturbed at its level, occurs at the different localities in the Drury and Hunua district, where coal is found. The average thickness of that coal seam may be estimated to amount to six feet. The section of the seam at Mr Fallwell's farm can be taken as a fair average.

The seam consists there of three portions, the upper part a laminated coal of inferior quality, one foot; then a band of shale, two inches; the middle part coal of a good quality, one and a half feet; then a band of bituminous shale, six inches; the lowest part coal of the best quality I have seen, two and a half feet. Thus the whole thickness of the coal itself may be considered to amount to about five feet. The bituminous shale accompanying the coal contains fossil plants, principally leaves of *Dicotyledones*. It is remarkable that no fossil ferns are found in connection with the Drury coal-beds; it is the more so, as at the other locality which I must mention—on the West Coast, seven miles from Waikato Heads—*only* fossil ferns, in a most beautiful state of preservation, are imbedded in gray argillaceous strata, alternating with sandstone and small coal seams of probably the same geological age as the Drury coal. A considerable number of specimens from both localities will, by a future examination, furnish the opportunity for determining the principal features of the flora of the *Brown Coal Period in New Zealand*.

The fossil gum found in the coal is a kind of "Retinite," derived from a coniferous tree, perhaps related to the Kauri; but it is by no means identical with the Kauri Gum, which is only found in the surface soil in those localities where there have been kauri forests. The fossil gum and kauri gum are very different in their qualities, as the most simple experiments in their ignition will show.

The thickness of the forest and the inaccessability of the country prevent our now ascertaining, in an exact manner, the extent of the Drury coal-field. Still, the existing openings show an extent of the coal-field quite large enough to encourage any Company to work the coal in an extensive manner.

I am glad to hear that a Company, under the name of "The Waihoihoi Mining and Coal Company" is formed, to begin the working of this coal.

The same kind of coal I saw again on the northern slope of *Taupiri* and *Hakarimata range*. At *Kupakupa*, on the left bank of the Waikato, I examined a beautiful seam about 150 feet above the level of the river. The thickness of the seam then exposed was about 15 feet; how much greater the thickness may be it is impossible to say, as the floor has never been uncovered.

This is the seam to which the attention of the inhabitants of Auckland was directed several years ago by my friend the Rev. A. G. Purchas. I believe several tons were at that time brought to Auckland ; but owing to various circumstances—the chief of which was the Native ownership—the hope of obtaining a supply from thence for Auckland was abandoned. No better position could, however, be found for mining purposes ; and the day cannot be far distant when it will be worked to supply fuel for the *steam navigation of the Waikato—the main artery of the Province of Auckland.*

I have reason to believe that a Coal Field of considerable extent exists on the borders of the wide plains on both sides of the Waikato, between Taupiri and Mangatawhiri—for which district, shut in on all sides by ranges, I propose the general geographical name of " *The Lower Waikato Basin.*"

A third coal-field exists on the Western and Southern boundaries of the very fertile alluvial plains above the junction of the Waipa and Waikato, which may be distinguished as " *The Middle Waikato Basin*"—the *future granary* of the Northern portion of this Island.

The localities in which coal has been discovered are the following :—in the Hohinipanga range, West of Karakariki on the Waipa ; near Mohoanui and Waitaiheke, in the Hauturu range on the upper branches of the Waipa ; and again in the Whawharua and Pareparo ranges on the Northern side of Rangitoto mountains.

THE NEW ZEALAND BROWN COAL.

The following are the results of several analyses of specimens of the *Drury Brown Coal*, sent to England some months ago by Mr Turnbull. The analyses have been forwarded to me by Mr Farmer :

Laboratory, Museum of Practical Geology,
Jermyn-st., London, April 13, 1859.

Sir,—I have completed the analyses of the coal (lignite) which you left at the Museum, and herewith furnish you with the results of the examination.

I am, Sir, yours obediently,

Brown, Esq   CHAS. TOOKEY.

Per-centage composition of Lignite, from Auckland.

Carbon	55·57
Hydrogen	4·13
Oxygen	15·67
Nitrogen	1·15
Sulphur	0·36
Ash	9·00
Water	14·12
	100·00

Coke .... 50·78 per cent.

The amount of sulphur is small, and this will be a point for favourable consideration in the application of the coal for smelting purposes. The whole of the water is expelled at a temperature of 120° centigrade.

Dundee Gas Works,
March 17, 1859.

Analysis of Auckland Coal.

Produce of gas per ton of coal carbonized, 9,632 cubic feet.

Illuminating power of gas, 1·75.

Durability : the length of time that a 4-inch jet requires to consume a cubic foot of gas, 53 minutes.

Specific gravity, 495.

Produce of coke per ton, carbonized, 9¼ cwt.

(Signed) JOHN Z. KAY,
Engineer Gas Company.

Gas Works, Berwick,
March 12, 1859.

New Zealand Coal.

Gas, in cubic feet, per ton of coal, 7,617.

Coke, per ton of coal, in lbs., 1,155.

Tar and ammoniacal liquor, per ton of coal, in lbs., 571.

Value of gas, per ton of coal, in lbs. of sperm, 384.

One cubic foot of gas, burned in a No. 2 fishtail burner (or union set), equal sperm candles, 3·12.

Value of one cubic foot of gas, in grains of sperm, 374·40.

Coke, trable, retains the granular structure of the coal; disintegrates when exposed to air; during combustion gives out little heat; and leaves a large mass of stone-coloured ash; specific gravity, 1·471.

$$\text{Composition} \begin{cases} \text{Combustible matter} & 39\cdot25 \\ \text{Silica and alumina} & 54\cdot44 \\ \text{Protoxide of iron} & 6\cdot31 \end{cases}$$

100·00

This coal is well adapted for the purpose of gas manufacture; the quantity produced is not large, but you will observe of a high quality, approaching several of the Scotch cannels in illuminating power.

The coke is of very inferior quality for heating purposes; but the quantity of iron share found in it is so great that it may possibly turn out to be a product of value.

JAMES PATTERSON,
Civil Engineer.

I subjoin comparative average analyses of three principal kinds of fuel, from which it may be seen that the Drury coal is precisely similar to the European brown coals in the proportion of its three principal constituents:

	Wood.		Brown Coal.		Black Coal and Anthracite.	
Carbon	51·4 to	52·6	55 to	76	73 to	96·51
Oxygen	43	42	26	19	23	3
Hydrogen	6	5·5	4·3	2·5	5·5	0·5

I embrace the opportunity of saying a few words on the *commercial value and applicability of the New Zealand Brown Coal.*

Although of entirely different character, and, generally speaking, of inferior value, to the older coals of the Primary formation, I cannot see any reason why this kind of coal should not be used in New Zealand for the same purpose as a similar brown coal is extensively applied to in various parts of Europe, and particularly in Germany, where it supplies the fuel for manufactures of all kinds, for locomotives and steamers, and for domestic purposes. I am perfectly familiar with this kind of coal, and can assure the people of Auckland, that the Brown Coal of this country is quite as good as that which is used in Germany for the purposes I have just mentioned. I would strongly recommend that any Company which may be formed for the purpose of working the coal should also at the same time establish *Potteries* for the manufacture of earthenware. Remarkably suitable *Clays* of every necessary variety have been shown to exist in the immediate neighbourhood of the coal-fields, by the borings which have been made by the Provincial Government at my request.* By the establishment of such works, the value of the coal would be made apparent to every-

---

* The following are the results of two borings made in the flats between Drury Hotel and the Drury Ranges, under the directions of Mr. Ninnis, to whom I am indebted for the tables subjoined:—

BORING No. I.

	Feet.	Inch.	
1)	2.	0	Dark soil.
2)	9.	6	Plastic clay, yellow and blue.
3)	1.	6	Gravel and pebbles.
4)	1.	0	Yellow clay.
5)	3.	0	Grey clay.
6)	6.	0	Blue clay.
7)	11.	0	Arenaceous clay.
8)	15.	0	Grey clay.
9)	2.	0	Greenish clay.
10)	1.	0	Dark grey clay.
11)	5.	0	Bluish grey clay.
12)	2.	0	Sandy clay.
13)	5.	2	Volcanic ashes and gravel.
14)	5.	6	Hard basaltic rock.
	69.	8	

BORING No. II.

	Feet.	Inch.	
1)	1.	0	Dark soil.
2)	7.	0	Yellow clay.
3)	6.	6	White clay.
4)	7.	0	Yellow and red clay
5)	1.	4	Brown clay.
6)	8.	0	Yellow clay.
7)	5.	0	Brown.
8)	4.	0	Reddish.
9)	10.	0	Brown.
10)	4.	6	Gravel and Volcanic ashes.
11)	9.	6	Hard basaltic rocks.
	63.	10	

Of these I would draw attention to No. I., 2, for common pottery; No. I., 6 and 8, for fine stoneware ; No. I., 7, for firebricks. The various coloured clays, No. II., 2 to 9, will be applicable to every kind of pottery. No. II., 8, may be used as a colour or pigment in the same way as ochre and umber are generally used.

body, and the manufacture itself, if properly conducted, cannot fail to be remunerative. It may be interesting to you to know that the far-famed "Bohemian Porcelain" is burned by means of brown-coal, from a seam of, in some places, 90 feet thickness. While stating the uses to which brown coal may be applied, I must warn you against thinking that it is suitable for steamers having to make long sea voyages. The bulky nature of the "brown-coal" will always prevent such steamers taking it on board when they can procure "black-coal." But, on the other hand, its qualities as a gas-producing coal, as the above analyses show, will render it valuable as an article of export.

I now come to another series of the older Tertiary strata, examples of which are found occurring in great regularity on the West Coast from Waikato to Kawhia. The lowest are argillaceous—the middle, calcareous—the upper, arenaceous.

The characteristics of the first *clayey strata* are, a light grey colour, very few fossils, small crystals of iron pyrites and glauconitic grains, which give these clay marls a similarity to the Gault and Green sands of the Cretaceous formation in Europe. They are found on the Eastern branches of Whaingaroa, Aotea, and Kawhia harbours.

Of greater interest and importance are the calcareous strata, consisting of tabular *limestone*, sometimes of a conglomerate nature, sometimes more crystalline, the whole mass of which is formed of fragments of shells, corals, and *foraminiferæ*, interspersed with perfect specimens of terebratulæ, oysters and pectens, and other shells. This limestone, when burned, makes excellent lime, and may be wrought and polished for architectural purposes.

The beds of Limestone worked by Messrs. Smith and Cooper, in the Wairoa district, belong to this formation, as do also the rich fossiliferous strata from the Waikato Heads towards Kawhia Harbour.

Picturesque columnar rocks of the same nature, looking almost as if they were artificially built of tabular blocks, adorn the entrance to Whaingaroa Harbour; and the romantic limestone scenery, and the fine Caves of the Rakaunui river—a branch of Kawhia Harbour—are deservedly prized by the settlers of Kawhia Harbour.

The Limestone Formation attains its greatest thickness (from 400 to 500 feet) in the Upper *Waipa* and *Mokau* district, between the Rangitoto range and the West Coast. It has in this country many remarkable features.

No one can enter without admiration the Stalactite Caves of *Tana-uri-uri* at Hangatiki, and of *Parianewanewa* near the sources of the Waipa—the former haunts of the gigantic *Moa*.

I went into those caves in the hope of meeting with a rich harvest of Moa skeletons, but I was sadly disappointed, those who had been before me in the days of Moa enthusiasm having carried off every vestige of a bone. Great, however, was my labour, and not little my satisfaction, in dragging out the head-less and leg-less skeleton of a Moa from beneath the dust and filth of an old raupo hut! The Maoris, seeing the greediness with which the "pakehas" hunted after old Moa bones, have long since carefully collected all they could find, and deposited them in some safe hiding-place—waiting for the opportunity of exchanging them for pieces of gold and silver, showing thus how well they have learned the lesson taught them by the example of the "pakeha."

The subterranean passages of the rivers in the *Pehiope* and *Mairoa* district are highly characteristic of the limestone formation. The limestone rocks, fissured and channeled, are penetrated by the water, and the streams run below the limestone upon the surface of the argillaceous strata, which I have before mentioned as underlying the limestone. This also explains the scarcity of water on the limestone plateau which divides the sources of the Waipa and Mokau rivers. The plateau is covered with a splendid growth of grass, and would form an excellent cattle run but for the deep funnel-shaped holes which everywhere abound. The Natives call them "*tomo.*" They are similar to the holes which occur in the limestone downs in England, and on the Karst mountains on the shores of the Adriatic Gulf, where they are called "*dolines.*"

The third and uppermost stratum of the older tertiary formation consists of beds of fine fossiliferous *sandstone*, in which quarries of good building stone may be found. There are whole ranges parallel to the primary mountains which seem to consist of this sandstone. I will mention only the *Tapui-*

whane range, about 2000 feet above the level of the sea, in which is the pass from the Mokau to the Whanganui country.

Without a map on a large scale, which I have had no time to prepare, it would be useless to enter more minutely now into a description of the various localities in which the different formations occur. I may, however, mention that limestone and brown-coal have been found in places to the North of Auckland, in the districts from Cape Rodney to the North Cape.

The horizontal beds of sandstone and marls which form the cliffs of the Waitemata, and extend in a Northerly direction towards Kawau, belong to a newer tertiary formation, and, instead of coal, have only thin layers of lignite. A characteristic feature of this *Auckland tertiary formation* is the existence of beds of volcanic ashes, which are here and there interstratified with the ordinary tertiary layers.

I must say no more on the tertiary sedimentary formations, in order that I may leave some time to devote to the *volcanic* formations, which, from their great extent and the remarkable and beautiful phænomena connected with them, render the Northern Island of New Zealand, and especially the Province of Auckland, one of the most interesting parts of the world.

VOLCANIC FORMATIONS AND PHENOMENA.

Lofty trachytic peaks covered with perpetual snow, a vast number of smaller volcanic cones presenting all the varied characteristics of volcanic systems, and a long line of boiling springs, fumaroles, and solfataras, present an almost unbounded field of interest, and, at the same time, succession of magnificent scenery.

It is only through a long series of volcanic eruptions, extending over the tertiary and post-tertiary periods, that the Northern Island has attained its present form. It would be a difficult task to point out the ancient form of the antipodean Archipelago the site of which is now occupied by the Islands of New Zealand. I must confine myself to a simple indication of the events which have given this country the form it was found to have by the South Sea Islanders on their arrival, many centuries ago, from the Samoan group—a form in all main respects the same as is now before our eyes.

The first volcanic eruptions were *submarine*, consisting of vast quantities of trachytic lava, breccia, tuff, obsidian, and pumicestone, which flowing over the bottom of the sea, formed an extensive submarine volcanic plateau. The volcanic action continuing, the whole mass was upheaved above the level of the sea, and new phenomena were developed. The eruptions going on in the air instead of under the sea, lofty cones of trachytic and phonolithic lava, of ashes and cinders, were gradually formed. These eruptions, breaking through the original submarine layers of trachytic lava, breccia, and tuff, raised them, and left them, as we now find them, forming a more or less regular belt round the central cones, and having a slight inclination from the centre outwards. These belts I shall have occasion to refer to under the name of " *tuff-craters*," or " *cones of tuffs*" or "*craters of elevation*." In the course of time the volcanic action decreased, and we must now imagine that tremendous earthquakes occurred—that parts of the newly formed crust gave way and fell in, forming vast chasms and fissures, which are now occupied by the lakes, hot springs, and solfataras.

Thus we now find in the central part of the Northern Island an extensive volcanic plateau of an elevation of 2,000 feet, from which rise two gigantic mountains, *Tongariro* and *Ruapahu*. They are surrounded by many smaller cones, as Pihanga, Kakaramea, Kaharua, Rangitukua, Puke Onake, Hauhanga. The natives have well named these latter, "the wives and children of the two giants Tongariro and Ruapahu;" and they have a legend to the effect, that a third giant, named *Taranaki*, formerly stood near these two—but quarrelling with his companions about their wives, was worsted in combat, and forced to fly to the West Coast, where he now stands in solitary grandeur, the magnificent snow-capped beacon of Mount Egmont (8,270). These are the three principal trachytic cones of the Northern Island.

By far the grandest and loftiest of the three is *Ruapahu*, whose truncated cone, standing on a basis of about 25 miles in diameter, attains a height of 9,000 to 10,000 feet above the level of the sea—about 3,000 feet of which is covered with glaciers and perpetual snow. Ruapahu, like Taranaki, is extinct. *Tongariro* alone can be said to be active. I was enabled to distinguish five craters on Tongariro, three of which are to a certain extent active. Steam is always issuing from them, and the natives state that from the principal crater, called *Ngauruhoe*, on the top of the highest cone of eruption (7,500), occasional eruptions of black ashes and dust take place, accompanied with loud subterranean noises. I may remark, that the shape of the cone is changing, the western side, for instance, having, during the great earthquake at Wellington, in 1854, fallen in, so that the interior of the crater is now visible from the higher points in the Tuhua district on the Upper Whanganui. The remarkable fact, that snow does not rest upon some of the upper points of the Tongariro system, while the lower ones are covered all the winter through, shows that those parts are of a high temperature.

I had no opportunity myself of ascending Tongariro, but I have met with the following interesting account of an ascent of the highest cone of eruption by Mr H. Dyson, which was communicated to the *New Zealander* by A. S. Thomson, M.D. :—

Mr. Dyson's Account of his Ascent of Tongariro.

"In the month of March, 1851, a little before sunrise, I commenced my ascent alone, from the north-western side of the Rotoaire lake. I crossed the plain and ascended the space to the northward of the Whanganui river. Here I got into a valley covered with large blocks of scoria, which made my progress very difficult. At the bottom of the valley runs the Whanganui river. After crossing the river, which at this place was then not more than a yard broad, I had to ascend the other side of the valley, which, from the unequal nature of the ground, was very tedious, and I kept onwards as straight as I could for the top of the mountain. At last I came to the base of the cone, around which there were large blocks of scoria which had evidently been vomited out of the crater, and had rolled down the cone. The most formidable part of my journey lay yet before me, namely, the ascent of the cone, and it appeared to me from the position where I stood that it composed nearly one-fourth of the total height of the mountain. I cannot say at what angle the cone lies, but I had to crawl up a considerable portion of it on my hands and feet, and as it is covered with loose cinders and ashes, I often slid down again several feet. There was no snow on the cone of the mountain, unless in some crevices to which the sun's rays did not penetrate. There was not on the cone any vegetation, not even the long wiry grass which grows in scanty patches up to the very base of the cone. The ascent of the cone took me, I should think, four hours at least; but as I had no watch, it is possible from the laborious occupation I was at, that the ascent of the cone looked longer than it was. But whether it was three or four hours I was clambering up the cone I recollect I hailed with delight the mouth of the great chimney up which I had been toiling. The sun had just begun to dip, and I thought it might be about 1 p.m., so that I had ascended the mountain from the Rotoaire lake in about eight hours. I must confess, as I had scarcely any food with me, that I kept pushing on at a good pace. On the top of Tongariro I expected to behold a magnificent prospect, but the day was now cloudy, and I could see no distance. The crater is nearly circular, and from afterwards measuring with the eye a piece of ground about the same size, I should think it was six hundred yards in diameter. The lip of the crater was sharp: outside there was almost nothing but loose cinders and ashes; inside of the crater there were large overhanging rocks of a pale yellow colour, evidently produced by the sublimation of sulphur. The lip of the crater is not of equal height all round, but I think I could have walked round it. The southern side is the highest, and the northern, where I stood, the lowest. There was no possible way of descending the crater. I stretched out my neck and looked down the fearful abyss which lay gaping before me, but my sight was obstructed by large clouds of steam or vapour, and I don't think I saw thirty feet down. I dropped into the crater several large stones, and it made me shudder to hear some of them rebounding as I supposed from rock to rock—of some of the stones thrown in I heard nothing. There was a low murmuring sound during the whole time I was at the top, such as you hear at the boiling springs at Rotomahana and Taupo, and which is not unlike the noise heard in a steam engine room when the engine is at work. There was no eruption of water or ashes during the time I was there, nor was there any appearance that there had been one lately. I saw no lava which had a recent appearance; notwithstanding all this, I did not feel comfortable where I stood in case of an eruption. The air was not cold—the ascent had made me hot—but I had time to cool, for I remained at the crater nearly an hour. At about 2 p.m. I commenced my descent by the same way that I ascended. A fog or cloud passed over where I was, and caused me to lose my way for a short time. When descending I saw between Tongariro and Ruapahu a lake about a mile in diameter. I could see no stream flowing out of it on the western side. An extinct crater may also be seen near the base of Tongariro. It was almost dark before I reached the Whanganui river, and, although in strong condition and a good walker, I felt completely done up, and I fell asleep in a dry water-course. The night was cold, but I slept soundly until daylight, when I immediately rose and continued my descent, and at 10 a.m. I reached my residence at Rotoaire, with the shoes almost torn off my feet."

As far as I can learn, Mr Dyson, in 1851, and Mr Bidwell, in 1839, are the only Europeans who have ascended the highest cone of Tongariro.

The difficulty of ascending Tongariro is still the same as when Dr Thomson published the foregoing account. "It does," he says, "not entirely arise from its height, or the roughness of the scoria, but from the hostility of the natives, who have made the mountain 'tapu,' or sacred, by calling it the backbone and head of their great ancestor. All travellers who have asked permission of the natives to ascend Tongariro, have met with indirect refusals. The only way to get over this difficulty is, to ascend the mountain unknown to the natives of the place, or even your own natives. Mr Dyson did this, but his ascent was discovered by a curious accident. During his progress up the mountain he took for a time the little frequented path which leads along the base of Tongariro to Whanganui. A native returning from that place observed his footmarks, and knew them to be those of a European. As he saw where the footsteps left the path, he, on his arrival at Rotoaire, proclaimed that a European was now wandering about alone on the sacred mountain of Tangariro. The natives immediately suspected it was Mr Dyson, and they went to his house, waited his return, and took several things from him. He was now a suspected man, and his conduct was watched."

The second active crater of the Tongariro system, at the top of a lower cone north of Ngauruhoe, is called *Ketetahi*. According to the natives, the first eruption of this crater took place simultaneously with the Wellington earthquake of 1854. From Taupo Lake I saw large and dense volumes of steam, larger than those from Ngauruhoe, emerging from the Ketetahi crater. The third active point on the Tongariro system is a great solfataraon the north-western slope of the range. The hot sulphurous springs of that solfatara are often visited by the natives on account of the relief they experience in respect to their cutaneous diseases.

A grand impression is made upon the traveller by those two magnificent volcanic cones—Ruapahu, shining with the brilliancy of perpetual snow—Tongariro, with its black cinder-cone capped with a rising cloud of white steam ;—the two majestic mountains standing side by side upon a barren desert of pumice (called by the natives, *One-tapu*), and the whole reflected, as by a mirror, by the waters of Lake Taupo.

LAKE TAUPO is about 28 English miles long, and 20 broad. This lake is surrounded by elevated pumice-stone plateaus, about 2000 feet above the sea, and 700 feet above the lake. The Waikato River, taking its rise from Tongariro, flows through the lake, traversing the pumice-stone plateaus on either side. In accordance with the names I have already proposed for the Middle and Lower Waikato Plains, the Taupo Country will form the "*Upper Waikato Basin.*"

It is one of the most characteristic features in the structure of the Northern Island, that, from the shores of Taupo Lake, an almost level pumice-stone plain—called *Kaingaroa Plain*—stretches at the foot of the East Cape range, with a very gradual descent to the coast between Whakatane and Matata—a plain which, though now presenting a sterile appearance, will, I hope, at no distant day, be converted into fine grassy plains, capable of supporting large flocks of sheep.

In a similar way, a higher volcanic plateau, consisting of trachytic tuff and breccia, and various other volcanic rocks, stretches in a more northerly direction to the East Coast, between Maketu and Tauranga, the farthest extremities of which reach even to the Auckland District. On one side of Hauraki Gulf, the Coromandel range is covered with trachytic breccia, and again, on the West Coast, the same rock forms the coast-range from *Manukau* to *Kaipara*. This extensive plateau is intersected by many deep valleys, the sides of which are characterised by a succession of remarkable terraces. The same plateau is also broken in many places by more or less regular trachytic cones from 1000 to 3000 feet high. That you may become acquainted with the geological character of such mountains, I will mention several examples, the names of which are well known amongst European settlers. To this class of mountains belong *Karioi* on the West Coast, near Whaingaroa, *Pirongia* on the Waipa, the regular cone of *Kakepuku* between the Waipa and the Waikato, *Maungatautari* on the Waikato, *Aroha* on the Waihou, *Putauaki* or Mount Edgecombe on the East Coast, and many others. The only active mountain which belongs to this class is *Whakari* or White Island, in the Bay of Plenty, a solfatara like the active crater of Tongariro.

Mr. David Burn, in his account of "A Trip to the East Cape," says:—

"In about an hour after passing Flat Island, the snowy vapour upon White Island began to be discernible. By 1 p.m. we were in immediate contiguity with this remarkable island, passing quite close to its southern extremity. As we made our gradual approach, its aspect was of the most singular description. Except on its northern point, to which the sulphurous vapour does not seem to reach, it is utterly destitute of vegetation; there are patches of growing underwood; but in every other direction, the island is bald, bleak, and furrowed into countless deep-worn ravines. After we had passed it a short distance to the eastward, the capacious basin of the crater, with its numerous geysers roaring and raging, exposed its sulphurous bosom to our eyes and nostrils. If the outer and western sides of White Island be blank and furrowed, its inner circle is chased, as it were, in a rare and picturesque manner—the sides of the hills, from their lofty mountain summits to the base, being *combed* into innumerable longitudinal ridges of a florescent bronze of brilliant and variegated hue.

"Of this island, Captain Drury, of H.M.S. Pandora, gives the following description in the 'New Zealand Pilot':—

"'White Island, or Whakari, is about three miles in circumference, and 860 feet high. The base of the crater is one and a half miles in circuit, and level with the sea. In the centre is a boiling spring about 100 yards in circumference, sending volumes of steam full two thousand feet high in calm weather. Around the edges of the crater are numberless smaller geysers sounding like so many high-pressure engines, and emitting steam with such velocity that a stone thrown into the vortex would immediately be shot in the air.

"'Here and there are lakes of sulphurous water, dormant; but the whole island is so heated as to make it difficult to walk. From the edges of the crater to the scene below is only to be compared to a well-dressed meadow of gorgeous green, with meandering streams feeding the boiling cauldron; but on approaching, we find this green to be the purest crystallised sulphur.

"'No animal or insect breathes on this island, scarcely a limpet on the stones, and 200 fathoms will hardly reach the bottom within half a mile of its shores.'

"Being under the lee of the island and in smooth water, Captain Bowden, in the most obliging manner, hove the steamer to, and lowering one of the quarter boats, conveyed us on shore to enjoy a personal inspection of this grand natural curiosity. There are two spots at which a landing may be effected, at the openings of the outer base of the crater; by a very little exertion in clearing away some of the boulders, the landing may be rendered perfectly easy; but although, this day, the water was smooth, still there was such a swell that judgment and caution were requisite to pick out a spot where best to escape the rollers that tumbled on the rough and broken beach.

"Never shall we forget the grand displays which we beheld in this sulphurous caldron. Its paintings fresh from Nature's hand—its lake of gorgeous green—its roaring jets of stormy vapour—are things to be witnessed, difficult to be described; but surpassing all these, and as if their central attraction, there was a fountain, seemingly of molten sulphur, in active play, which shot a column of wide-spreading green and gold into the scorching atmosphere. The beauty of this fountain was surpassing, and we were under the impression, that from its energy, the volcano was more than commonly active in its workings. We were very circumspect in our approaches, as the surface in places was soft and yielding, and we knew not to what brimstone depths an unwary step might sink us. Our difficulty in walking, therefore, arose less from the heat, though that in places was great, than from the apprehension of sinking too far in the soft crustaceous surface, from which diminutive spouts of vapour would spit forth as if to resent our intrusion. Whenever we thought the ground at all doubtful, we sounded our way by hurling large stones to see what impression they would make, and we adventured or avoided proceeding accordingly.

"Time, to our great regret, would not admit of a minute exploration, but all the grand features of the island had passed under view. We looked in vain for the gorgeous meadow described by Capt. Drury; but we had only to enlarge any of the numberless miniature vapour holes to obtain pure crystallised sulphur *hot from the bakery*, and at the same time to convert these holes into more active vapour jets. The streams that issued in various directions were of boiling heat, limpid and tasteless; but, though sulphur was everywhere strewn around, it did not appear to be in quantities sufficient for shipment. After an hour's stop, we returned to our ship greatly delighted with the visit, and much indebted to our obliging captain for having put it in our power to enjoy it."

Mr. Heaphy has kindly furnished me with a map and views of this singularly interesting island.

If we take a wider view of the geological features and the physical outline of these just described high plains and plateaus consisting of regular layers of trachytic rocks, breccia, and tuff, we shall find that the steep cones of Ruapahu and Tongariro rise from the centre of a vast tuff cone of extremely gradual inclination, the basis of which occupies the whole country from shore to shore—from East to West—having a diameter of 100 sea miles, and forming the largest cone of *tuffs*, or in other words, the largest *crater of elevation*, in the whole world.

The Hot Springs.

Intimately acquainted with the described volcanic phenomena of the active and extinct volcanic mountains, are the *Solfataras*, *Fumaroles*, and *Hot Springs*. They are found in a long series, stretching across the country in a N.N.E. direction, from the active crater Ngauruhoe in the Tongariro system, to the active crater of White Island (Whakari). They occupy the chasms and fissures to which I have already referred.

There is only one other place in the world in which such a number of hot springs are found that have periodical outbursts of boiling water—that is, in *Iceland*, the well-known *geysers* of which are of precisely similar character to those in New Zealand. The geysers or boiling fountains of Iceland, long celebrated for possessing this property in an extraordinary degree, have, indeed, strong rivals in the *puias* and *ngawhas* of New Zealand. Although there may be no single intermittent spring in New Zealand of equal magnitude with the great geyser in Iceland, yet in the extent of country in which such springs occur, in the immense number of them, and in the beauty and extent of the siliceous incrustations and deposits, New Zealand far exceeds Iceland.

In enumerating the principal of this phenomena, we may begin with—

1. The active craters of *Tongariro*, which are at present in the condition of solfataras that may be called the state of repose of active craters, and with the hot springs rising on the slope and at the base of that mountain.

2. We then pass on to the *Tokanu* and *Terapa* springs on the Southern extremity of the Taupo lake. The principal "puia" at Tokanu is called *Pirori*, an intermittent fountain whose column of boiling water, of two feet in diameter, sometimes reaches a height of more than 40 feet.

3. On the opposite side of Taupo, at the Northern extremity of the lake, we again meet with hot springs, and with a river of warm water called *Waipahihi*, which, rising in the extinct volcanic cone of Tauhara, falls, in a vapour-crowned cascade, into Taupo.

4. Descending from Taupo by the outlet of the Waikato, we find, on the left bank, in the midst of a great number of pools of

boiling mud, a fumarole called *Karapiti*, an enormous jet of high-pressure steam, escaping with such force as to produce a sound like letting-off the steam from huge boilers, and as to eject to a great height sticks, or the like, thrown in by the curious traveller. On the right bank is another fumarole of similar character, called *Parakiri*.

5. About twenty-five miles below the outlet of the Waikato from Taupo, at the "pa" *Orakei-korako*, both banks of the rapidly-flowing river are perforated, in more than a hundred different places, by fumaroles and boiling springs, most of which are of the intermittent kind; and siliceous incrustations of beautiful colours decorate the banks of the river. *Temini-a-Homaiterangi*—the principal geyser—throws up its large column of boiling water at intervals of about two hours to a height from 20 to 30 feet. An immense volume of steam succeeds the jet, and the water then suddenly sinks into the basin.

6. At Orakei-korako the line of hot springs crosses the Waikato, and continues along the foot of the very remarkable *Pairoa range* on the Easterly side of the Waikato. The almost perpendicular Western side of this range is caused by an immense "fault" in the volcanic plateau, corresponding to a deep fissure in the earth-crust, from which sulphureous acid, sulphuretted hydrogen, sulphur and steam, are continually escaping, while huge bubbles of ash-coloured mud are rising on the surface.

7. From the same range, the warm-water river *Waikite* takes its origin. On both sides are deep pools of boiling water, on the margins of which we discovered most beautiful ferns, hitherto unknown, one species belonging to the genus *Nephrolepis*, the other to the genus *Goneopteris*. These ferns are remarkable not only for their elegance, but also from the peculiar circumstances under which they exist, as they are always surrounded by an atmosphere of steam.

8. We now come to the well-known ROTOMAHANA, the most wonderful of all the wonders of the Hot Springs district of New Zealand. I will not attempt to describe in a hasty lecture like this the beauties of this Fairy-land. Whoever has had the happiness to look into the blue eyes of *Otukapuarangi* and *Te Tarata* can ever forget their charms? and whoever has stood

beside the boiling surf of the *Ngahapu* basin will always retain a vivid impression of its terrors. The terrace of siliceous deposit on the shores of Rotomahana are unequalled in the world, nor is there anything that even bears any resemblance to them.

9. On the *Roturua* lake the intermittent boiling springs of *Whakarewarewa* are the most interesting. *Waikite*, the principal "ngawha," issues from the top of a siliceous cone some 20 feet high, and is surrounded by several smaller geysers, boiling mud-pools, and solfataras. At intervals of considerable length, sometimes extending to many months, all these "*ngawhas*" begin to play together, and form a scene which must be most wonderful and beautiful.

The hot springs of *Ohinemutu* form agreeable bathing places, the fame of which is already established.

10. The last in the line are the great solfataras on the pumice-stone plateau between Rotorua and Rotoiti—such as *Tikitere* and *Ruahine*.

I will now say a few words in explanation of these phenomena.

All the waters of the Springs are derived from atmospheric moisture, which, falling on the high volcanic plateau, permeates the surface and sinks into fissures. Taupo—the axis of which corresponds with the line of the Hot Springs—may also be considered as a vast reservoir, from which the lower springs are supplied. The water, sinking into the fissures, becomes heated by the still-existing volcanic fires. High-pressure steam is thus generated, which, together with the volcanic gases, decompose the trachytic rocks. The soluble substances are thus removed by the water, which is forced up, by the expansive force of the steam and by hydrostatic pressure, in the shape of boiling springs. The insoluble substances form a residuum of white or red fumarole clay, of which the hills at Terapa round Rotomahana and the Pairoa consist.

All the New Zealand hot springs, like those of Iceland, abound in Silica, and are to be divided into two distinct classes —the one *alkaline*, and the other *acid*. To the latter belong the solfataras characterised by deposits of sulphur, and never forming intermittent fountains. All the intermittent springs belong to the alkaline class, in which are also included the most of the ordinary boiling springs. Sulphurets of Sodium and Potassium, and Carbonates of Potash and Soda, are the solvents of the Silica, which, on the cooling and evaporation of the water, is deposited in such quantities as to form a striking characteristic in the appearance of these springs.

Here I must leave this interesting subject. To enter more deeply into the theory of these phenomena would be out of place here. It may be, however, well to mention that numerous facts prove that the action which gives rise to the hot springs is slowly diminishing.

I must also state my conviction that ere long these hot springs will be visited by many travellers, not only for the sake of their beauty and interest, but also for the medicinal virtues they have been proved to possess. Already many Europeans have bathed in, and derived benefit from, the warm waters at Orakei-korako and Rotomahana.

I am unwilling to omit the interesting legend current among the Natives in reference to the origin of these hot springs. The legend, as told by Te Heuheu, the great chief on the Taupo lake, is the following :—

The great chief *Ngatiroirangi*, after his arrival at Maketu at the time of the immigration of the Maoris from Hawaiki, set off with his slave Ngaurunoe to visit the interior, and, in order to obtain a better view of the country, they ascended the highest peak of Tongariro. Here they suffered severely from cold, and the Chief shouted to his sisters on Whakari (White Island) to send him some fire. This they did. They sent on the sacred fire they brought from Hawaiki, by the taniwha *Pupu* and *Te Haeata*, through a subterranean passage to the top of Tongariro. The fire arrived just in time to save the life of the Chief, but poor Ngauruhoe was dead when the Chief turned to give him the fire. On this account the hole through which the fire made its appearance—the active crater of Tongariro—is called to this day by the name of the slave *Ngauruhoe*; and the sacred fire still burns within the whole underground passage along which it was carried from Whakari to Tongariro.

This legend affords a remarkable instance of the accurate

observation of the Natives, who have thus indicated the true line of the chief volcanic action in this island.

Having now described the older and more extensive volcanic phenomena of *the interior*, I proceed to notice the later phenomena of volcanic action in the *immediate neighbourhood of Auckland*.

THE AUCKLAND VOLCANIC DISTRICT.

The isthmus of Auckland is completely perforated by volcanic action, and presents a large number of true volcanic hills, which, although extinct and of small size, are perfect models of volcanic mountains. These hills—once the funnels out of which torrents of burning lava were vomited forth, and afterwards the strongholds of savage cannibals—are now the ornaments of a happy land, the home of peaceful settlers, whose fruitful gardens and smiling fields derive their fertility from the substances long ago thrown up from the fiery bowels of the earth.

My Geological Map of the Auckland District contains no less than sixty points of volcanic eruption within a radius of *ten miles*—the variety of which, together with the regularity of their formations, gives very great interest to this neighbourhood. The newer volcanic hills around Auckland are distinguished from the older ones in the interior, not only by their age, but by the different character of their lava—the older being *trachytic*, while the Auckland are all *basaltic*. I have not yet mentioned the difference between Trachyte and Basalt. I will therefore say a few words in explanation. The difference consists in the minerals of which the rocks are composed. Trachyte is composed of a mixture of glassy feldspar (*Sanidin*) and hornblende: obsidian and pumice-stone are the usual concomitants of trachytic lava. Basalt consists of a minutely-crystalline mass of feldspar mixed with augit; an admixture of greenish grains of Olivin is characteristic of basalt.

In order to gain a clear idea of the history of the Auckland volcanoes, we must suppose that before the period in which the Auckland isthmus was slowly raised above the level of the sea, a submarine volcanic action was already going on. The products of this submarine action are regular beds of volcanic ashes, which form highly interesting circular basins with strata always inclining from within, outwards. You will at once remember several striking examples which I can mention—as the Pupuki Lake on the North Shore; Orakei Bay in the Waitemata; Geddes's Basin (*Hopua*) at Onehunga; and the tidal Basin (*Waimagoia*) at Panmure;—Pupuki Lake, believed to be bottomless, has been ascertained by Captain Burgess (who kindly sounded it at my request) to be only 28 fathoms. I call those basins and similar formations, *tuff-craters* or *tuff-cones*. The excellence of the soil of Onehunga and Otahuhu is owing to the abundance of such formations, decomposed strata of which form the richest soil that can be met with. It is curious to observe how the shrewder among the settlers, without any geological knowledge, have picked out these tuff craters for themselves, while those with less acute powers of observation have quietly sat down upon the cold tertiary clays.

After the submarine formation of the tuff-craters, the volcanic action continuing, the isthmus of Auckland was slowly raised above the sea, and then the more recent eruptions took place, by which the cones of scoria, like Mount Eden, Mount Wellington, One Tree Hill, Mount Smart, Mount Albert, and Rangitoto, were formed and great out-flowings of lava took place. Many peculiar circumstances, however, prove that those mountains have not been burning all simultaneously. It can easily be observed that some lava streams are of an older date than others. In general the scoria cones rise from the centre of the tuff-craters (Three Kings, Waitomokia, Pigeon Hill near Howick.) Occasionally, as in the instance of Mount Wellington, they break through the margin of the tuff-crater.

The Crater System of Mount Wellington is one of the most interesting in this neighbourhood, as beautifully shown by the large map which Mr Heaphy has kindly prepared for me from actual survey. There are craters and cones of evidently different ages. The result of the earliest submarine eruptions is a tuff-crater. The Panmure road passes through the tuff-crater, and the cutting through its brim exhibits beautifully the characteristic outward inclination of the beds of ashes, elevated from their former horizontal levels by the eruptions, which threw up the two minor crater cones south of the road—one of which is now cut into by a scoria quarry. After a comparatively long period of quiescence, there arose from the margin of the first crater system the great scoria cone of Mount Wellington, from whose three craters large streams of basaltic lava flowed out in a Westerly direction, extending North and South along the existing valleys of the country, one stream flowing into the old tuff-crater, and spreading round the bases of the smaller crater cones. The larger masses of these streams flowed in a South-westerly direction towards the Manukau, coming into contact with the older and long before hardened lava streams of " One Tree Hill." The traveller on the Great South Road will observe about one mile East of the "Harp of Erin Inn" the peculiar difference in the colour on the road, suddenly changing from red to black, where the road leaves the older and more decomposed lava streams of One Tree Hill and passes on to the new and undecomposed lava streams of Mount Wellington. The farmers have been able to avail themselves of the decomposed lava surface, which is now beautifully grass covered, but not of the stonefield of the newer Mount Wellington and Mount Smart streams.

The *Caves* at the "Three Kings," Pukaki, Mount Smart, Mount Wellington, &c., are the result of great bubbles in the lava streams—occasioned probably by the generation of gases and vapour as the hot mass rolled onward over marshy plains. These bubbles broke down on their thinnest part—the roof—and the way into the caves is always directly downwards.

Examples of every gradation may be seen—from the simple tuff-crater without any cone, to those which are entirely filled up by the scoria cones. Especially interesting are those which may be said to represent the middle state, in which there is a small cone standing like an island in a large tuff-crater, and surrounded by either water or swamp. Perhaps the most perfect specimens of this kind occur at Otahuhu and near Captain Haultain's, a map of which, from actual measurement, has been prepared by Mr. W. Boulton. You need not be alarmed when I tell you, that even the very spot on which we are assembled is the centre of an old tuff-crater, from which fiery streams once issued, and which has thrown out its ashes towards the hill on which the barracks stand.—In order to account for these various shapes, it must be borne in mind that the cones of scoria were once higher, but on the cessation of volcanic action they sunk down in cooling, and some entirely disappeared.

That the Auckland volcanoes were, in the true sense of the word, "burning mountains," is proved not only by the lava-streams, which are immense in comparison to the size of the cones, but also from the pear-shape volcanic bombs which, ejected from the mountain in a fluid state, have received their shape from their rotatory motion through theair. That the eruptions of the Auckland volcanoes have been of comparatively recent date, is shown by the fact that the ashes everywhere occupy the surface, and that the lava-streams have taken the course of the existing valleys. This is beautifully exemplified by the probably simultaneous lava streams of Mount Eden, the Three Kings, and Mount Albert, which, flowing through a contracted valley, met altogether—on the Great North Road— and form *one* large stream to the shore of the Waitemata, terminating on the well-known long reef West of the Sentinel Rock. But many thousand years have passed since Rangitoto, which is probably the most recent of the Auckland volcanoes, was in an active state.

I have been frequently asked whether it is true, as a countryman of mine who some years ago travelled in New Zealand is said to have told the European settlers, that New Zealand is a pleasant country, but that they had come a thousand years too soon. In answer to this I have to remark that any one who knows anything of geological science must be aware, that "a thousand years" is an almost inappreciable space of time in reference to geological changes. And I would rather say, that it would have been better for New Zealand if it had been colonized a thousand years ago, as there would have then been no cause for the discussion of the "Land Question."

I should have much pleasure in saying a great deal more on the Geology of New Zealand, but time will not permit me. Many subjects I have been compelled to omit altogether—such as the quartary formation in the Drury, Papakura, and Waiuku flats; the Basaltic Boulder formation; the Alluvial formations in the Middle and Lower Waikato Basin, and other places; and I have said nothing of the changes which are now going on.

The materials which I have accumulated during my six months' sojourn in New Zealand will, I expect, require several years of labour to prepare for publication; and unless the war which now threatens my own country should unhappily interfere to prevent the completion of the peaceful scientific undertaking of the Expedition to which I belong, it will give me great pleasure to forward to Auckland copies of our publications respecting New Zealand, accompanied by an atlas, containing the maps and other illustrations.

In concluding this lecture, I cannot omit the opportunity of saying a few words of farewell to the inhabitants of this Province.

Now that I am on the point of leaving Auckland, I turn in memory to the hour in which I made up my mind to leave my friends on board the Frigate "Novara" and to remain for a while in New Zealand. I can assure you it was an hour of great anxiety, but I am glad to say I have never regretted the decision to which I with so much difficulty brought myself. Having received assistance in my labours from all sides, I have arrived at results which have afforded me much satisfaction, and which I hope will not be without good fruit to the present and future inhabitants of this Province. Having at first felt some difficulty in making up my mind to remain, I now feel a similar difficulty in leaving. Home ties, however, are drawing me homewards, and I must quit the country in which I have spent so many happy days. In parting, I have one request to make—that you will remember me as kindly as I will remember you; and I have one wish—which is for the prosperity of the colony of New Zealand, and the advancement of the Province of Auckland.

EXPLANATION OF THE MAPS.
By Dr. Ferdinand v. Hochstetter.
MAP I.
NEW ZEALAND.—GEOGRAPHICAL AND GEOLOGICAL SURVEY.

NEW ZEALAND consists of two large and several small Islands, which form a broad strip of land extending from the South-West to the North-East, and at its Northern end prolonged by a narrow Peninsula in a North-Westerly direction. The outlines are very similar to those of Italy in a reversed position. Its geographical position also harmonises with that of Italy, being situated between the parallel circles of $34\frac{1}{2}°$ and $47\frac{1}{2}°$ Southern width, and the meridians of $166\frac{1}{2}°$ and $178\frac{3}{4}°$ Eastern length of Greenwich. Its length is 800 sea miles, its central width from East to West is 120 sea miles (30 German miles), and the area of the whole group of Islands amounts to 99,969 English square miles. New Zealand is therefore nearly as large as Great Britain and Ireland.

Two Straits—Cook's Straits in the North, and Foveaux Straits in the South—separate New Zealand into three parts of different sizes—two larger Islands, which, in the absence of other names, have been termed the North and South Islands, and a small Isle called Stewart's Island. To these the first English Governor, Captain Hobson, officially gave the names of New Ulster, New Munster, and New Leinster (after the three Provinces of Ireland). These names sometimes figure on the maps, but are only remembered by the colonist as antiquated reminiscences. The original name of New Zealand is Te Ika a Maui—that is, the Fish of Maui (Cook wrote Ea heino Mauwe)—a name which has a mythical signification. Also Te Wahi Punāmu, or land of the green-stone; and Ra Kiura. The former was applied only to the South Island, where the mineral Nephrite, which was so highly prized by the Maoris, was to be found.

The three Islands form a geological group, being parts of the same system, which forms one distinct line of elevation in the Pacific Ocean. And Nature, with her mighty forces of fire and water, has indelibly engraved the history of the Islands on their surface. In the South, wild alpine regions covered with ice and glaciers, and in the North, volcanoes reaching to the regions of eternal snow, are seen glimmering in the distance by the mariner on approaching the coast. The fertile, richly-watered alluvial flats are the virgin soil on which the settler forms his new home, and where, blessed with the most salubrious of all climates, he has to combat only the wilderness to ensure the reward of his labour.

The characteristic of New Zealand is a large longitudinal mountain chain, which, broken by Cook's Straits, runs through the principal Island in a South-Westerly and North-Easterly direction from the South Cape to the East Cape. This forms the backbone of the Islands, and reaches its grand and multifarious development in the Southern Island, where it assumes, in numberless summits covered with snow and glaciers, the character of mountains, to which, with full justice, the name of the Southern Alps has been given. Majestically in the centre of these mountainous regions stands the summit of Mount Cook, with its neighbouring giant heights, elevated 13,000 feet above the level of the sea, or nearly the height of Mont Blanc. Mighty glaciers, streams, and magnificent mountain lakes, splendid cascades, passes, and dark clefts whose rocky walls re-echo the noisy torrents rushing through them, form the beauty of a wild solitary mountain scenery, seldom trodden by human feet. The brave explorers who have of late years had the courage to penetrate into these wild regions,* report that their grandeur if even equalled is not excelled by any in the world.

Towards the West, those Alpine mountains abruptly assume a very precipitous character, and form, on this the stormy side of the Island, a dreadfully rugged, weather-beaten, and rocky coast. On the East, at the feet of these mountains, lay wide-spreading plains and alluvial flats, well adapted for agriculture, and which are occupied by the European settlers as sheep runs; while on the North and South the gradations and slopes of the mountains are of a clay slate formation, in which are hidden those quartz veins that have of late years been developed into the rich gold-fields to which Nelson and Otago owe their prosperity.

In the North Island, past Cook's Straits, the Southern Alps have their continuation in the great mountain chain which

---

* Mr. Julius Haast, the German traveller, geologist to the Province of Canterbury, deserves the highest tribute of praise for his researches in the Southern Alps. In 1860 he investigated the mountain ranges of Nelson, and in 1860 and 1862 those of the Province of Canterbury, where he reached the highest central summit of Mount Cook, and discovered here numerous glaciers to about 3,000 to 4,000 feet above the level of the sea, while the height of the eternal snow region commences at 7,500 to 8,000 above that level. The principal glaciers Haast named Clyde, Havelock, Ashburton, Godley, Murchison, Tasman, Hochstetter, Müller, Hooker; while the principal summits are called Mount Tyndall, Mount Forbes, Mount Arrowsmith, Mount Petermann, Mount De la Beche, Haidenger Range, Malte Brun Range, Mount Elie de Beaumont, &c., &c.,—all mountains of 10,000 feet and upwards.

stretches along the East Coast, from Cape Pallisser to the East Cape. Different peaks, which have names, such as Tararua, Ruahine, Kaimanawa, Te Waiti, are of pretty nearly equal height. The highest summits reach only from 5,000 to 6,000 feet, and are therefore much less than the height of the Southern Alps. These mountains are an almost *terra incognita*, and doubtless contain treasures of many kinds. The Northern Island is also rich in Volcanic phenomona. The high plateau on the Western side of this mountain chain, sloping off towards the North and South, forms the remaining part of the Island, and is pierced to a great depth in more than a hundred places by Volcanic agencies.

High Trachytic Volcanoes, and a great number of small Basaltic eruption-cones, of quite a recent age, and a long chain of hot springs which, like the Geysers of Iceland, at intermittent periods, throw up masses of boiling water in steaming fountains, Fumaroles and Solfataras in a multitude of forms of the utmost conceivable grandeur, offer to the geologist a rich field for research, and to the traveller some of the most remarkable scenes of nature.

The extraordinarily diversified surface formation of New Zealand leads to the inference of a most varied geological conformation. The commencement of a geological examination of the North and South Islands has proved this to the fullest extent, during the last few years. The geologic detail maps of my own observations, and partly those of my friend Haast's, show manifold changes in strata and in minerals. They show that, throughout the whole chain, from the oldest metamorphic formation to the latest sediment layers, and also from the earliest plutonic rocks, up to the latest volcanic formation, all the principal genera are represented.

New Zealand is rich in minerals of all kinds, and all those which are now found—as gold, copper, iron, chrome, graphite and coal—can only be regarded as the first-fruits of future treasures to be brought to light in years to come.

The fossil fauna and flora of New Zealand, as far as at present known, differs entirely from that of Australia, and many geological facts prove that New Zealand, surrounded by the ocean, has been an island—though not in its present form—since the most remote ages, and entirely isolated from larger continents. In the later ages of the history of man, it has been inhabited near the coast and along the course of the larger rivers. It has thus maintained the peculiar and pristine origin of its fauna and flora. The European domestic animals which have only been introduced since the latter end of the last, and the commencement of the present century, are the only quadrupeds which existed in this country.

In respect of its insular position, its splendid oceanic climate, the fertility of its soil, and its entire formation, New Zealand is of all the Colonies of the British Crown, the most similar to the Home Country, and destined to become the mother of a new civilized race—a Great Britain of the South Sea.

MAP II.
THE GEOLOGICAL FORMATION OF THE SOUTHERN PART OF THE PROVINCE OF AUCKLAND.

Since Dieffenbach's memorable travels in New Zealand in the year 1840, no naturalist has visited the southern part of the Province of Auckland, so justly celebrated for its grand volcanic phenomena.

The geological information which Dieffenbach gave in his work could not suffice; the interior was topographically almost unknown. A journey, therefore, to these parts, promised rich field for observation; and after a stay of nearly two months in Auckland and its neighbourhood, I started with a numerous suite, well provided for a long journey on foot, and for a campaign in that thinly inhabited country.

On the 6th of March, 1859, near Maungatautari, I reached the main stream of the Waikato, flowing from the heart of the Island.

I travelled along this river in the canoes of the natives, and observed near Kupa Kupa large brown coal seams, and entering the Waipa, visited the Mission Station, and took a tour to the West to visit the Harbours of Whaingaroa, Aotea, and Kawhia. All those localities are of geological importance on account of the numerous localities for petrefactions. At Kawhia, I found, besides Belemnites, the first Ammonites found in New Zealand.

From Kawhia, I took a tour inland to the Mokau district. Penetrating through numerous primitive forests, and traversing large mountain chains, I passed the springs of the Wanganui River in the Tuhua district, and on the 14th of April, our party arrived at the majestic Lake Taupo, which is surrounded by the grandest volcanic scenery, and is situated 1250 feet above the level of the sea. Here I was in the heart of the country, at the foot of the steaming volcano, Tongariro, and its now silent neighbour Ruapahu, 9,200 feet high, covered with eternal snow. On the Southern side of the Lake is a Mission Station, where I received the kindest hospitality, and my Maori companions were entertained in Maori fashion in the neighbouring Pa Pukawa by the great chief Te Heuheu. After I had sketched the plan of the Lake, and examined the numerous hot springs on its borders, I started from the sources of the Waikato River flowing from that Lake, and followed the most interesting chain of boiling springs, solfataras, and fumaroles, which are situated in a North-Easterly direction between the active crater of Tongariro and the island volcano of Whakari or White Island, on the East Coast. The Lake neighbourhood is situated in the line where the *ngawhas* and *puias* of New Zealand (that is, the boiling fountains and geysers, where silicious stalactites form terraces of basins) reach their most magnificent development near the Rotoiti and Rotomahana Lakes. I consider the hot springs in this district the most remarkable, and, next to Iceland, the largest in the world.

In the beginning of May, I reached the East Coast near Maketu, Tauranga Harbour. Hence I went inland to the Waiho valley, or the valley of the Thames of New Zealand, and arrived at Maungatautari again at the Waikato. I wandered through the fruitful fields of the Middle Waikato basin, by Rangiawhia, the centre of the Maori settlement, and paid a visit to the Maori King Potatau Te Wherowhero, at his residence Ngaruawahia, at the confluence of the Waikato and Waipa, and returned by the Waikato, to Maungatawhiri, at the end of May, on my way to Auckland.

The result of this expedition, extending over three months, was in every respect satisfactory; the favourable state of the weather lessened many of the difficulties which travelling through swampy rivers and the almost impassable New Zealand bush would otherwise have entailed, and luckily my journey happened to be during the harvest of potatoes, wheat, &c., &c., consequently I found no want in the commissariat department. We received the most hearty welcome at the various Mission Stations scattered over the country, and the native Chiefs everywhere received the Te Rata Hokiteta (my name in the Maori tongue,) and my companions, at their pas, with great honour and hospitality, were always willing to oblige, and with good-humoured zeal ready to assist with everything.

With the assistance of my friends Haast, Hay, Koch, and Hamel, who were my companions, the results proved in every way as satisfactory as could be expected. A rich collection of Geographical, Mineralogical, Botanical, and Geological observations came into my hands, and also for Ethnographical studies I had ample opportunity.

My principal aim was, however, the Geography and Geology of the country. To make geological sketches, I was obliged to work at the same time topographically, as the maps of the interior were based only on the reports of travelling missionaries and *a-la-vue* sketches.

The map which I took with me from Auckland, for my guidance, only gave a little information about the coast, and its value beyond a few miles from Auckland was not more than that of a piece of waste paper. I had, therefore, to adopt a triangulation system based on the nautical observations of Captain Drury, and carried out from the West to the East coast, with the energetic assistance of Major Drummond Hay. The natives, who otherwise always manifested their mistrust of the Government Surveyors, and placed every difficulty in their way, did not interrupt me. They knew that I was a foreigner who would remain only a short time in the country, and they assisted me in every possible manner, to enable me to relate in my distant land much of the beauties of their country. The chiefs themselves were my guides to the most interesting parts, and to the top of the mountains, where with the utmost readiness they gave me the names of the mountains, rivers, valleys, and lakes, and explained after their fashion the geography of the district. I carefully collected all the names they gave me, and trust that I have rescued from oblivion many beautiful Maori names. The terrain positions I sketched always on the spot, and in this manner I returned with materials from which I compiled in Auckland a map of the Southern part of that Province, on a large scale. *

This map has been re-cast and revised with the assistance of my original sketches and surveys, by Dr. A. Petermann, and the map in this Atlas, in the reduced scale, is the product of his labour, and the result of my observations. It stands to reason, that a map which contains nearly 2,500 miles (10,000 square miles) and embraces more than the fourth part of the Northern Island, executed by the assistance of a compass alone, within the period of three months, can make no pretensions to a trigonometric exactness. It is, however, the first map which gives a correct view of the rivers and mountain systems, and of the lakes in the interior of the Northern Island, and will be useful until some better and more complete map takes its place. The Barometrical measures which I took served as corresponding observations to those of the Observatory of the Royal Engineers in Auckland, which were kindly placed at my services by Colonel Mould.

The geological condition of the Southern part of the Province of Auckland may be sketched in the following order:—

PALÆOZOIC (PRIMARY) FORMATION.

Dark coloured claystone, old sandstone called grauwacke, silicious and jasparoid slate, form a complex system of layers,

---

* A copy of my original map, to the scale of 2 miles to 1 inch, remained in Auckland for the use of the Government. A second copy was sent to Mr. J. Arrowsmith, in London, to be used for the construction of a large New Zealand map in six parts, which that gentleman intended to compile, with the understanding, however, that this map was to be used only as a provisional delineation of my observations. The Geological map of the Province of Auckland, which was exhibited in the International Exhibition of London, in 1862, by Mr. Charles Heaphy, was entirely a copy and combination of my maps and surveys, without any acknowledgment of my authorship. The map, also, of the Isthmus of Auckland, given in the Quarterly Journal of the Geological Society of London, by Mr. Charles Heaphy, was published without my knowledge, and is a very incomplete copy of my observations and maps, which were in Mr. Heaphy's official charge. In this map that gentleman also introduced his own observations upon the geological formations of the neighbourhood of Auckland, made previous to my arrival in New Zealand, but without possessing even the most elementary knowledge necessary for making a Geological Survey. I have felt it my duty to make these remarks out of respect for truth and science.

which on the Northern Island, where crystalline (metamorphic) slate rocks have not yet been discovered, appears to be the oldest formation, the geological age of which it is impossible to state with exactitude, as petrifactions have not yet been discovered. The gold quartz veins which are to be found in the peninsula of Cape Colville (Coromandel harbour), are imbedded in this old clay slate formation. Frequent outbursts and layers of dioritic rocks correspond to the Silurian age.

Distribution.—Upon the peninsula of Cape Colville, where it is covered to a great extent with recent volcanic conglomerate, containing rich gold quartz veins, which have given rise to mining enterprise since 1862 (Coromandel Gold Fields). The alluvial diggings arising from those veins are of little importance. Upon the islands of the Hauraki Gulf, where the Great Barrier and Kawau contain veins of copper ore (copper pyrites, black and some red copper ore). Upon Waiheki are immense strata of jaspar and petrosilex; as also in the mountain chains of the west side of the Firth of the Thames, and thence in a southerly direction into the chains of Taupiri and Kakarimata. Further on they are covered by tertiary and volcanic strata, and penetrate to the surface only in a few localities. The mountain chains which extend from the Wellington district to the Taupo Lake and the East Coast, consist most likely also of old palæozoic rocks.

MESOZOIC (SECONDARY) FORMATION.

The Mesozoic formation has been ascertained by the discovery of Ammonites and Belemnites in the entrance to the Waikato river, and in the harbour of Kawhia, although the exact age of these strata cannot be fixed by the discovery of petrifactions up to this time, and which are enumerated as follows:—

1. A very large complex stratum of very regular and highly inclined beds of marl and sandstone, on the Waikato South Head, with

    Belemnites Aucklandicus (v. Hauer),
    Aucella plicata (Zittel),
    Placunopsis striatula (Zittel),
    Terebratula spec.;

At the Kawhia Harbour, with

    Belemnites Aucklandicus (var. minor),

Ammonites Novo-Seelandicus (v. Hauer),
Inoceramus Haasti (Hochst.).

2. Strata containing coal on the West Coast, south of the entrance of the Waikato—sandstone, marl, and slate clay, with thin, worthless seams of coal, and numerous portions of plants, amongst which are frequently to be found in good preservation:

Polypodium Hochstetteri (Unger),
Asplenium palæopteris (Unger);

while the Belemnites (belonging to the group of Canaliculati) indicate the Jurassic system. The largely folded Inoceramus and Ammonites have a greater similarity to those from the chalk formation.

CAINOZOIC (TERTIARY) FORMATION.

Tertiary Formations are distributed over a large portion of the Province of Auckland, for the most part in a horizontal position.

1. *Brown Coal Formation : sandstone and clay slate, with beds of useful coal.*

   (*a.*) The Hunua coal-field, near Drury and Papakura district, south of Auckland, discovered in the year 1858 by the Rev. Mr. Purchas, and worked since 1859 by the Waihoihoi Coal Company. The coal belongs to a class of brown coal—to the so-called glanz and pitch coal—and contains a fossil gum—Ambrit (Haidinger)—which has often been mistaken for kauri gum. The price of this coal in Auckland is 30s. to 32s. per ton. The argillaceous slate and sandstone accompanying this contains several bivalves and leaves of Dicotyledones:

   Fagus Ninnisiana (Unger),
   Lorantophyllum Griselinia (Unger),
         ,,   Dubium (Unger)
   Myrtifolium lingua (Unger) &c.

   (*b.*) The coal-fields of the Lower Waikato basin—a large brown coal basin—is situated at Kupakupa, on the northern declivity of Hakarimata chain, but is not yet worked.

(c.) Brown coal strata on the west and southern brim of the central Waikato basin.

2. *Marine argillaceous marl, sand, and limestone, with numerous petrifactions.*

   (a.) Waitemata beds: sandstone and marl, on the isthmus of Auckland, on the North Shore and Manukau, with stray pieces of wood transformed into brown coal.

   On the Orakei Bay, near Auckland, strata rich in glauconite, many foraminiferæ and bryozoæ with small pectens—

   Pecten Aucklandicus (Zittel),
   ,,   Fischeri (Zittel)—

   small forms similar to Bivalves and Belemnites, which are most probably the centres of Vaginella shells.

   (b.) The limestone cliffs of Drury, near Auckland: flat limestone rich in foraminiferæ, with Turbinolia, Schizaster, Terebratula, Pecten, &c.

   (c.) Waikato Heads and southerly direction of the West Coast: granulated sandstone strata, resting irregularly on the above-mentioned Belemnite beds with coal layers, and containing Cidaris, Nucleolites, Schizaster, Fasciculipora, Retepora, Cellepora, Waldheima, Pecten, Sharks' teeth, &c.

   (d.) Clay marl and flat sandstone on the borders of Whaingaroa, Aotea, and Kawhia, on the West Coast, with Pecten, Waldheimia, &c., and many Foraminiferes.

   (e.) The flat coarse limestone in the Upper Waipa, Maungapu, and Mokau district, with many subterrestial rivulets, caves, and funnel-shapel holes.

POST-TERTIARY (OR QUATERNARY) FORMATION.

1. Plastic clay and sand, with Lignite in the Lower Waikato basin, and in the flats on the south and east side of the Manukau harbour.

2. The *terrace* formation in the Lower and Middle Waikato basin—the terraces the number and regularity of which causes astonishment to the observer, are the consequences of the continual erosions of the rivers during a slowly continuing rising of the lands in the quaternary period.

The Taupo district is the source of the extensive masses of pumice which are distributed over the terraces.

3. *Littoral formation along the coast.*
   (*a.*) Formation of downs mostly near the West Coast and on the Coast of the Bay of Plenty.
   (*b.*) Layers of titanic magnetic iron sand along the West Coast.
   (*c.*) Mud with brackish sea animals in the estuaries of the East and West Coast.

4. *Formation in the interior.*
   (*a.*) Extensive swamps and peat bogs along the East Coast, the Middle and Lower Waikato basin, and on the entrance of the Waikato.
   (*b.*) Layers, with bones of the Moa (*Dinornis*), and Moa stones in swamps; alluvial deposits and caves in the Upper Waipa, Mokau, and Tuhua district, and on the East Coast.
   (*c.*) Layers of kauri gum in the northern part of the Province of Auckland, where formerly kauri forest existed.
   (*d.*) Alluvium containing gold in the neighbourhood of Coromandel harbour.

5. Accumulations by the hands of men, as shells, stones, bones, &c., over different districts.
   (*a.*) Heaps of shells of edible varieties—Cardium, Ostrea, Mytilus, Patella, Venus, Haliotis, Mesodesma, Turbo, Monodonta, &c.—particularly in the places of former pas and villages, analogous to the Kjokken moddings in Denmark.
   (*b.*) Stones of fire-places of the Maoris, charcoal, and ashes.
   (*c.*) A variety of tools made from stone by the Maoris, anchors, axes, &c., prepared of Aphanite, Nephrite, flintstone, &c.
   (*d.*) Human bones, bones of dogs, whales, fishes, and different birds—Penguin, Albatross, Weka (*Ocydromus*), Kiwi (*Apterix*), Moa bones, and egg shells, in the neighbourhood of the Maori fire-places. These bones are mostly burned, and bear the marks of stone weapons.

VOLCANIC FORMATION.

1. *The north side of the Manukau Harbour* is formed of a rugged rocky coast wall composed of mighty layers of volcanic stone masses, consisting of angular fragments of the different volcanic basis of rocks—Trachy-Dolerite, Andesite, &c., which are transformed landwards into different coloured conglomerate clays.
2. *On the southern side of the Manukau Harbour*, and on both sides of the Waikato, thence to the Aotea Harbour, extensive strata of basaltic conglomerate cover the tertiary layers, and with these conglomerates are immediately connected masses of eruptive basalt, without forming distinct craters.
3. The *volcanic zone* which encloses the Middle Waikato basin, and is situated between this and the Lake Taupo, is principally formed by trachyte and pumice, with which are connected a very long line of volcanos, such as Karioi, Pirongia, Kakepuku, Maungatautari, Aroha, and many others. These mountains consist of trachytic, andesitic, and doleritic rocks; their summits are decayed and destroyed, and their craters scarcely recognisable..
4. The *volcanic formation of the Taupo zone* consists of a rhyolithic and trachytic lava. The volcanic eruption which commenced in the tertiary period continued, and gave to the Northern Island its present form only in the quaternary period. The eruptive masses of the Taupo zone consist of lava (the richest known) of silicious earth, also of *rhyolithic rocks* of all kinds, with obsidian and pumice. Near the centre of the Northern Island, on the southern border of the great inland lake Taupo, the water of which fills a large sunken crater, there rises on a plateau of pumice of 2,000 feet above the level of the sea the two giant volcanoes of New Zealand, Tongariro and Ruapahu. The Tongariro volcano, which rises to an elevation of 6,500 feet, is yet active as a Solfatara, with two large and constantly steaming craters (Ngauruhoe and Ketetahi). Ruapahu, which is 9,000 feet high, is covered with everlasting snow, and its fires seem to be extinct. These two mountains are accompanied on their northern side by a number of smaller extinct volcanoes,

which the natives have designated as the wives and children of the two giants.

In a north-easterly direction, a few miles distant from the coast in the Bay of Plenty, is situated the second active volcano of New Zealand, Whakari (White Island), 863 feet high, from the crater of which ascends, uninterruptedly, large white clouds of vapour. The distance between these two large volcanos amounts to 120 miles, and between them the volcanic agency steams and boils in more than a thousand places from deep furrows and fissures, a sign of the continual subterraneous fire, while numerous lakes are formed by the sunken ground, and which represents the Lake District, so celebrated for its boiling springs, fumaroles, and solfataras; or, as the natives call them, the *ngawhas* and *puias*, in the southern parts of the Province of Auckland (*vide* expl. Map 4).

5. The volcanic formation of Auckland zone is of basaltic lava (*vide* expl. Map 3).

RECORD OF THE HEIGHTS OF THE SOUTHERN PART OF THE PROVINCE OF AUCKLAND.

† Parts which have been adjusted by the marine survey ("New Zealand Pilot," and English charts).
b Barometric measures, by Hochstetter.
* Estimated.

English feet.

Auckland—Meteorologic Observatory of the Royal
    Engineer Department ... ... 140†
    Claremont House, upper end of Princes-street 130*b*

Kaipara Harbour, West Coast:
    Te Karanga Mountain, on the River Otamotea 1440†
    Wakakuranga, mountain on the Oruawharu
      River ... ... 476†
    Opara, Mount ... ... 378†
    Auckland Peak, by Otau Creek ... ... 1023†
    Koharanga, on the Kaipara River ... ... 326†

Titirangi chain, between the Waitakeri and the
    Manakau Harbour:
    Mount Tea Wekatuku ... ... 1430†

EXPLANATION OF THE MAPS.

	English feet.
Pukematiku, Henderson's Bush	1300*
Maungatoetoe, Delworth's Farm	1200*
Parera, West Coast	700*

Manakau Harbour, West Coast:

North Head, Paratutai Island, Signal Station...	350†
Pilot Station	300*
Pukehuhu	690†
Omanawanui Peak	1100*
Te Kaamoki, or Te Komoki Peak, near the Huia ...	480†
The Huia Peak	1280†
Puponga, highest point	390†
Heights of the left border of the big Muddy Creek...	600†
Heights by Whau Creek	800†
South Head, Mahauihaui	580†

East Coast, from the Bay of Islands to the Waitemata Harbour, or the Harbour of Auckland:

Cape Tewara, or Bream Head, Wangarei Harbour	1502†
Summit between Bream Head, near Wangarei...	1340†
Moto Tiri Island } Hen and Chickens {	725†
Taranga Island	1353†
Mount Hamilton, near Rodney's Point	1050†
Kawau Island, Mount Taylor	510†
Little Barrier Island, or (Houturu) Mount Many Peaks	2383†
Great Barrier Island, or (Aotea) Mount Hobson	2330†

The Volcanos of Auckland:

Rangitoto	920†
North Head, Takapuna	216†
Mount Victoria, Takarunga	280†
Heaphy Hill	100*
Mount Eden	642†
„ Hobson	430†
„ St. John	400†
„ Albert	400†
„ Kennedy	310†
„ Three Kings	390†

		English feet.
One Tree Hill	... ...	580†
Mount Smart	... ...	300*
„ Wellington	...	350*
Pigeon Hill		110†
Otara Hill	...	150†
Mangere Hill	...	333†
Waitomokia	...	120*
Puketutu		263†
Otuataua		300†
Maungataketake ...		300†
Maumrewa		300†
Matakarua	... ...	300†
Drury—Young's Inn (first storey)	...	75*b*
Brown coal shaft on Farmer's land ...	...	356*b*

Great South Road, between Drury and Mangatawhiri:

1. First hill at the entrance of the bush   491*b*
2. Highest point of the road 811*b*
3. Waikohowheke, house on the road   ...   598*b*
4. Second height of the road on the place where the view of the Waikato opens   ...   770*b*

Mangatawhiri	... ...	77*b*

Papahorahora, near Kupakupa, on the left side of the Waikato River, the place of the brown coal seam 250*

Taupiri, a hill on the right border of the Waikato, opposite the Mission Station 983*b*

Kakepuku, isolated mountain not far from the Mission Station on the Waipa 1531*b*

Points between the Waipa River and the West Coast:

  Toketoke, lake on the way from Whatawhata to Whaingaroa 249*b*
  Highest point of the road from Whatawhata to Whaingaroa 853*b*
  Whaingaroa Harbour, Captain Johnson's house   93*b*
  Station between the Whaingaroa and Aotea Harbours 243*b*
  Mill on the Oparau River, Kawhia Harbour ...   97*b*

EXPLANATION OF THE MAPS. 59

	English feet.
Pirongia, highest point of the Oparau River to the Waipa	1585*b*
Pirongia, highest point of the mountain group	2830†

Waikato River:

1. Near Mangatawhiri	35*
2. Near Rangiriri, pa on the right border of the Waikato	51*b*
3. Near Taipouri, island in the river with a Maori village	63*b*
4. Near Tukopoto, Mission Station	75*b*
5. Ngaruawahia, residence of the Maori King	85*
6. Kirikiriroa	97*b*
7. Aniwhaniwha, Waikato Bridge	166*b*
8. Near Orakei Korako	970*b*
9. Near the efflux from Taupo lake	1250*b*

Between the Waikato and Waipa:

| Maungatautari, Maori pa | 621*b* |
| Otawhao, Mission Station, | 211*b* |

Waipa River and District:

Ngaruawahia	85*
Whatawhata, left side of the Waipa	109*b*
School-house	112*b*
Kaipiha, Mr. Turner's house	167*b*
Waipa, at the entrance of the Mangaweka	143*b*
Mission Station of the Rev. A. Reid, nearly 25 feet above the bed of the river	173*b*
Awatoitoi, Maori settlement on the right border of the Waipa, nearly 25 feet above the bed of the river	185*
Orahiri, on the left bank of the Waipa	186*
Haugataki, Maori settlement	195*b*
Te Ana Uriuri, cave of Stalactites	201*b*
Tauahuhu, Maori settlement on the left bank of the Wangapu	196*b*
Mangawhitikau, Maori village	237*b*
Puke Aruhe, hill	877*b*

Upper Mokau District:

| Takapau, Maori settlement | 823*b* |

English feet.

Piopio, Maori settlement on the upper Mokau
River... ... ... 469*b*
Mokau River, above the Wairere falls ... 420*b*
Pukewhau, Maori pa on the left bank of the
Mokau River 683*b*
Mokauiti, between the Maori settlement Huritu
and Punanga 473*b*
Puhanga 937*b*
Morotawha, place of encampment on the 7th and
8th of April, 1859 ... ... 570*b*
Tarewatu mountain ridge, height of pass from
the Mokau to the Wanganui district ... 1581*b*
Tarewatu, highest point 1790*b*
Tapuiwahine, highest point on the way from
Makau to Wanganui 1933*b*

Upper Wanganui, Tuhua District:
Ohura, Maori village 917*b*
Katiaho, Maori settlement, on the Ongaruhe
river 650*b*
Ngariha, hill on the Ongaruhe river 1551*b*
Pokomotu plateau, highest point on the way from
Katiaho to Petiano 1386*b*
Petania, Maori village on the Taringamotu river 754*b*
Takaputiraha chain, passage from Petania to
Taupo 1534*b*
Pungapunga brook, on the road to Taupo ... 897*b*
Puketapu, mount on the road to Taupo ... 2073*b*

Lake Taupo:
Moerangi, pumice-stone plateau on the west and
south-west of Lake Taupo 2188*b*
Whakairomu 2175*b*
Kuratao river on the road to Pukawa ... 1719*b*
Poaru, Maori settlement 2289*b*
Pukawa, pa on the southern bank of Taupo
Lake 1399*b*
Mission Station of the Rev. Mr. Grace, at Lake
Taupo 1473*b*
Koroiti plateau, on the south bank of Lake Taupo 1768*b*
Taupo Lake (after Dieffenbach, 1337 feet) ... 1250*b*

	English feet.
Roto Aira, after Dieffenbach	1709
Rotu Punamu, after Dieffenbach	2147

Tongariro and Ruapahu:

	English feet.
Tongariro, Ngauruhoe Mount (after Dieffenbach, 6,200)	6500*
Ruapahu, on Taylor's map	10,236*
„ Arrowsmith's map	9000*
„ English charts	9195*
Pihanga	3500*

Between Taupo Lake and the East Coast:

	English feet.
Oruanui, Maori settlement	1672b
Plateau above Orakei Korako	2200*
Orakei Koroko, pa on the left bank of the Waikato river	1169b
Boiling mud springs at the foot of Paeroa	1409b
Waikite, hot springs at the foot of the Paeroa chain	1241b
Pakaraka, above Roto Kakahi	1801b
Roto Kakahi Lake	1378b
Roto Mahana Lake	1088b
Tarawera Lake	1075b
Papawera plateau, between Roto Mahana and Tarawera	1867b
Mission Station on Tarawera Lake, Rev. Mr. Spencer	1502b
Rotorua Lake	1043b
Ngongotaha, mount on the southern bank of Roturua	2282b
Rotokawa, small lake on the eastern bank of Rotorua	1098b
Waiohewa, or Ngae, settlement on the north-eastern bank of Rotorua	1103b
Pukeko, on the Rotoiti	1063b
Omatuku, near Maketu	1388b

East Coast:

	English feet.
Major Island (Tuhua), highest point	410†
Monganui mountain, at the entrance of the Tauranga harbour	860†
Plate Island (Motonau), centrum	166†

	English feet.
Whale Island, or Motu Hora, highest point ...	1107†
White Island, or Whakari...	863†
Mount Edgcumbe, eastern summit	2575†
East Cape (East Cape Islet)	420†

Between the East Coast and the Waiho River:

Waipapa brook, on the coast from Tauranga to the Waiho	803*b*
Heights of the Wanga chain, near the Wairere falls	1414*b*
Wairere river, immediately above the highest falls :..	1442*b*
Height of the pass over the Whanga chain, near the Wairere falls	1481*b*
The height of the Waiwere Falls	670*b*
Waiho Flats, near Wairere Falls	573*b*
Whatiwhati, settlement at the foot of the Patetere plateau	537*b*
Castle Hill (Cape Colville chain) near Coromandel Harbour	1610†

MAP III.

THE ISTHMUS OF AUCKLAND AND ITS EXTINCT VOLCANOS.

THE great southern part of the Northern Island of New Zealand is connected by a small isthmus with the northwestern peninsula, on the parallel circle of 37° S. lat. The sea penetrates through the Hauraki Gulf on the Eastern Coast, forming many branching creeks, and washes in a southwesterly direction into the so-called Waitemata River upon the north side of the peninsula. On the West Coast—the stormy weather side of New Zealand—the ocean, penetrating through hard volcanic rocks in a narrow entrance, spreads out and forms the Manukau basin, the southern coast of this isthmus. The land between the two seas is only some five or six miles broad, and in two places where the Waitemata River forms small creeks in a southerly direction towards the Manukau basin it narrows to the width of one mile. These narrow strips have been used by the Maoris from ancient times to carry their canoes from one side of the island to the other, and this has given rise to the idea of forming a canal, and thus uniting the two seas.*

While the Waitemata is the most central amongst the many harbours of the East Coast, the Manukau basin is, of all the harbours of the West Coast, the best, and the only one which is available, without danger, for large ships. The clear-sightedness of Captain Hobson, who, in 1840, recommended to the British Government, as the most suitable situation for the Capital of New Zealand, this isthmus, on both coasts of which navigation is so easy and safe, and which is so centrally and favourably situated for connecting both larger continents of the Northern Island, deserves all commendation. No other position in the Northern Island offers, by its central situation, the advantages of so easy and safe a water communication in all directions. Besides the numerous arms of the sea which penetrate in manifold directions deep into the land, there are numerous navigable rivers such as the Kaipara, Wairoa, Waikato, Piako, and Waiho, all of which are of easy access from this isthmus.

Auckland, the present capital of New Zealand, and the principal town of the Province of Auckland, is the seat of the Colonial and Provincial Governments, and was established in the year 1840. It is situated on the northern side of the isthmus, on the banks of the Waitemata. Of rapid growth, and extending itself from year to year, in 1861 this city numbered 8,000 inhabitants. More than this number occupy the vicinity of the town and the country of the district of Auckland.

A beautiful macadamised road leads from Auckland to Onehunga, or the Manukau Harbour. Onehunga—originally a settlement of pensioners, who each received from the Government a cottage and an acre of land—has rapidly progressed, and is now a considerable town, and on account of its pleasant position it is the favourite retreat of many of the wealthier class of Auckland. Between Auckland and One-

---

* The western isthmus is the so-called Whau portage, which is one mile wide, the highest elevation being 111 feet. The eastern isthmus is the Tamaki portage, near Otahuhu, south of Mount Richmond, and is only 3,900 feet long and 66 feet high.

hunga are numerous country seats, villas, and farms. Along the road, villages are rapidly forming, such as Newmarket, Mount St. John, and Epsom. Every sign of the former wildness of the isthmus has vanished. The old New Zealand vegetation has given way to European plants. Scoria walls and green hawthorn hedges divide the various estates; green meadows, gardens, and fields charm the eye. Everywhere herds of fine cattle are seen grazing in the fields. Omnibuses are constantly passing on the roads, and the whole forms a picture of a fresh and happy life.

The isthmus of Auckland is also one of the most interesting volcanic districts of the globe. It owes its distinguishing feature to a great number of extinct volcanos, with more or less distinctly preserved craters and lava streams which form extensive scoria fields at the foot of the hills, or with tuff-craters which encircle the scoria cones like an artificial wall, and are irregularly distributed over the isthmus and the neighbouring banks of the Waitemata and Manukau. The volcanic activity at each new eruption seems to have taken a different course from the former, and divided itself into numerous small cones. My map of the isthmus, which extends over a district of 20 miles in length, by 12 in width, shows not less than 63 independent points of eruption.

These are volcanos on the smallest scale, forming cones of an elevation of from 300 to 600 feet above the level of the sea. The highest amongst them is Rangitoto, which rises at the entrance of the harbour of Auckland to the height of 900 feet. But they are perfect models of volcanic cones and crater formation, and offer a large field of geognostic observation, refuting entirely the theory of elevation craters by Leopold von Buch.*

These mountains rise on a base consisting of tertiary sandstone and argillaceous marl, the horizontal and only locally disturbed strata of which are easily recognisable on the steep banks of the Waitemata and Manukau Harbours. The examination of these isolated points of eruption gives proof

---

* A description in detail will be given in the scientific publication of the "Novara" expedition, and will appear in the volume which embraces the geology of New Zealand.

of repeated and different volcanic outbursts in one and the same locality.

The first eruptions—probably sub-marine, at the bottom of a well-sheltered bay—consisted of loose masses, and of ruins of the fundamental basis, scoria and ashes. These eruptions took place in many shocks, following each other; the masses thrown out formed layers above each other and around the place of outbreak, causing a flat rising cone, with more or less circular or basin-like crater in the centre—tuff-cone and tuff-crater. The Pupuki Lake on the North Shore, the Orakei Bay, east of Auckland; Geddes' Basin, near Onehunga; the basin Waimagoia, near Panmure; and Kohuora Hills south of Otahuhu, are amongst other distinguished examples of such tuff-craters. Like the Maren in the Eifel, these crater basins are very deep and are filled with water. The sweet water lake, Pupuki, has a depth of 28·fathoms, or 168 feet. * They are sometimes flat, dry, or swampy. When they are situated near the sea it has generally forced an entrance, and ebbs and flows in and out of the crater basin. In consequence of their rich and fertile volcanic soil these tuff-cones hold an important position in the Province of Auckland—almost everyone of them is occupied by the homestead of a settler. The practical shrewdness of these men has led them, without geological knowledge, to settle at the basis or side of these craters—their flourishing meadows and clover fields contrasting strongly with the fern and manuka scrub (*Leptospermum*) of the clay soil.

With the beginning of the volcanic activity seems to have commenced, although very gradually, a rise of the whole isthmus; so that the later eruptions took place above the sea. In this second period, the volcanic activity increased to the emissions of red hot masses of scoria and streams of lava. At that time the Auckland volcanos were fire-spitting mountains in the true sense of the word; their steep cones at a slope of 30° to 35° were formed of scoria, volcanic bombs, and lapilles (Mount Eden, Three Kings, Mount Smart, Mount Wellington, and many others), with deep, funnel-shaped craters, and where

---

* It is the opinion of the translator that the lake is connected by a submarine channel with Rangitoto, which is the source of the lake.

repeated eruptions followed each other out of one and the same crater, cones of lava were formed again, like Rangitoto. Where these new eruptions followed the former course, new scoria cones grew up within the ring of the tuff-crater, and according to the number of the eruptions, or the sinkings which followed the extinct volcanic activity, larger or smaller islets were formed within, where water or swamp filled the tuff-craters. The lava of all the Auckland volcanoes is petrographically identical. It consists of porous basalt lava, rich in olivin, which makes a good building stone for the substantial erections in Auckland, while the scoria cones afford an excellent material for the roads of the isthmus.

The name of Rangitoto, which signifies "Sky of Blood," would lead to the supposition that the Natives have given this name in consequence of the reflection of the burning streams of lava in the nightly sky, and that therefore the Auckland volcanos have been in activity in very recent historical times; but this is improbable. That their activity belongs to the most recent geological period of the earth, and to the geological chronology of the present time, is proved by the fact that the volcanic ashes cover the surface directly, and that the lava streams have run by no means at one and the same time into the neighbouring valleys. These have therefore existed at the time of the emission of lava, and the surface of the district has since that time undergone no material change.

Transformed through the diligence and enterprise of the European settler into fertile cultivated districts, the Auckland volcanos are but monuments of a remarkable history of the Maori race. Only a few generations have passed since the Auckland isthmus was the seat of a mighty Maori tribe—the Ngatiwatuas—consisting of 20,000 to 30,000 men. These extinct fire mountains, with their commanding situations and wide prospects, occupied at that time, the position of hill forts, like the feudal castles of Germany. On their summits were the fortified pas of the chiefs, while at the foot of the hills were distributed the huts and kumera cultivations of the slaves. The slopes of the hills were formed into regular terraces, and fortified with palisades. The huts and houses are now destroyed; the palisades have disappeared; the Maori feudal castles have decayed; the terraces and holes are the only remaining monuments of a brave people which were annihilated in the bloody, cannibal wars of Hongi, the "Napoleon of New Zealand," in the years 1820 to 1830, and whose deeds live only in song and tradition.

MAP IV.

ROTO-MAHANA (OR THE WARM LAKE) AND ITS HOT SPRINGS.

The Lake District, so called on account of its numerous lakes, is situated about two days' journey from the Bay of Plenty. It is almost exclusively inhabited by the natives, who have selected the beautiful and fertile banks of Rotorua and Tarawera as their settlements. The Mission Station at Temu (the Rev. Mr. Spencer's residence) is at present the only European habitation, and is the resort of many travellers and naturalists, who visit the neighbourhood during the summer months. The principal point of attraction of this region is Roto-mahana, or the Hot Lake, with its wonders, a visit to which well repays the fatigues of a few days' travelling through New Zealand rush and swamps.*

It is one of the smallest lakes of the district, scarcely exceeding in length three-quarters of a mile from north to south, and in width a quarter of a mile. I hardly believe that this small, dull-green lake, with its swampy borders, and the surrounding barren and miserable-looking hills, which are destitute of trees, and only covered with fern, would come up to the expectation of the traveller, who has heard so much of its wonders. That which makes it the most remarkable of all the lakes of New Zealand, nay even the most remarkable of all spots of the earth, lies mostly hidden from the view of the new arrival—except the immense clouds of steam which rise everywhere—which leads to the supposition that in reality nothing is to be seen.

---

* The journey from Auckland is generally made in from one to two days by sea to Tauranga with a favourable wind. From Tauranga one can arrive in two days at Tarawera and Roto-mahana Lake, either direct or by Maketu—both roads equally bad. The return can be made over the Patetere plateau to the Waikato River, and from this by canoe to Mangatawhiri, whence the Great South Road leads to Auckland.

The name of "Warm Lake" (Roto—lake; mahana—warm) may in the full sense of the word be given to it. The masses of boiling hot water which spring up along the banks and from the bottom of the lake, are really collossal. Of course the whole lake is warmed by them, but the temperature of the water differs considerably in various places, as they are nearer or further from the springs. At many points, even in the centre of the lake, the thermometer rises from 30° to 40° c., (86° to 104° F.) while near its stream I found it only 26° c. (78·8° F.) The water is thick and swampy, and neither fish nor shell-fish can live in it. Otherwise the lake is a favourite resort of innumerable aquatic birds, who build their nests on its warm banks, while they find their food in the waters and swamps of the cold lake Roto-makariri. The natives shoot them at certain seasons, but at other times they do not permit either Europeans or themselves the pleasure of sport. The birds of Roto-mahana are at this period strictly "tapu."

Visitors who intend to stay a few days at the lake are recommended by the natives to select as their quarters the small island Puai. This is a rock, 12 feet high, 250 feet long, and nearly 100 feet wide. Small huts are there erected, in which we made ourselves as comfortable as possible. But I believe that any one who did not know that persons have lived here for several weeks, would only with great difficulty be persuaded to remain here even for one night. The continual roaring, rushing, singing, buzzing, boiling sound, and the intense heat of the ground, impresses a feeling of terror, and during the first night of my stay I awoke suddenly, as the ground under me became so hot that I could not possibly bear it. In examining the temperature, I made a hole in the soft ground, and placed the thermometer in it. It rose immediately to boiling-point, and when I took it out, a stream of hot steam instantly ascended; so that I hastened to cover it again as fast as I could. Indeed, the whole island is nothing but a torn and fractured rock, decomposed and softened by steam and gases, which, almost boiled to softness, may at any moment tumble to pieces, and vanish in the hot water of the lake. Hot water bubbles up everywhere, either below the surface of the lake or above it; and wherever a hole is made in the ground, or the crust removed which is formed over the fissures of the rock, hot steam bursts forth, which we used for cooking our potatoes and meat, spreading them on ferns, according to native custom.

The centre of attraction and of interest is the eastern bank, where are the most important of the springs, which indeed the lake has to thank for its renown, and which are the most magnificent and grand of all hot springs at present known.

Te Tarata is situated at the north-eastern end of the lake. It lies 80 feet above the level of the lake, within a crater which is open towards the side of the lake, and forms the principal basis of this mighty bubbling spring. It is 80 feet long by 60 feet broad, and filled up to the brim with clear boiling water, which issues in the centre several feet higher, looking beautifully blue in its snow-white incrusted basin. Enormous clouds of steam, rising upwards, are reflected in the blue mirror of the basin. The temperature of the water, which probably reaches to boiling-point in the centre, was 84° c. (183·2° F.) near the rim of the basin. The water is neither alkaline nor acid; it has a slightly salt taste, and possesses in a high degree the property of petrifaction, or rather of incrustation. The sediment consists, as in the hot springs of Iceland, of silica, and the overflow has formed on the slope of the hill a system of crystal terraces, which, appearing almost as white as marble, present a sight which it is impossible to describe. It is as if a cascade, rushing over steps, had been suddenly arrested, and transformed into stone. Each of these steps has a small elevated rim, from which hang delicate stalactites; and here and there, on the smaller and broader steps, are formed water basins. These blue basins, filled with crystal water, form natural baths, which could not be surpassed by those constructed by the most refined luxury. One can select his bathing-place either deep or shallow, small or large, and of every temperature according to his taste, as the basins situated on the heights near the source contain warmer water than those of the lower steps. Some of the basins are so large that a person can swim in them with comfort. Such is a description of the celebrated Te Tarata spring. The natives assert that the whole water in the principal basin is sometimes ejected suddenly with vast force, and that it is possible to look into the empty basin, thirty feet deep, which fills again speedily.

A path leads from the foot of the Te Terata spring through the bush to the great Ngahapu spring. The basin of this spring is 40 feet long and 30 feet broad. The water within it is in constant and dreadful agitation. It is only for a few moments that the water is quiet in the cauldron, when it again bubbles up, and is thrown eight to ten feet high; and a foaming surf of boiling hot waves stream over the walls of the basin; so that the observer is obliged timidly to retreat. The thermometer rises in these springs to 98° c. (208·4° F.) Further south, close to the banks, is situated the Te Takapo spring —a boiling water basin of 10 feet in diameter, the geyser eruption of which rises to a height of 30 to 40 feet.

Not far from this spring the traveller arrives at a hollow called Waikanapanapa (Variable Water), the approach to which is covered with bush, and somewhat difficult, as one has to pass several suspicious-looking places, where there is danger of sinking in the boiling mud. The cavity itself appears like the crater of a volcano; the walls, bare of vegetation, are rent and torn; pieces and tongues of rock of white, red, and &blue fumarolic clay rising upwards like spectres, threaten to fall every moment. The bottom is formed of fine mud, and silicious stalactites, broken into every form and variety, lie about like pieces of ice after the breaking up of a frozen stream. Here is a deep pool filled with bubbling mud—there a cauldron full of boiling water—near it a dreadful hole which, with a hissing noise, ejects a column of steam; and further on small mud hills (fumaroles), from two to five feet in height—mud volcanos, if the name may be applied to them—which, with a dull noise, throw out of their craters boiling mud, and represent, on a small scale, the effects of large volcanos. In the back-ground is situated a green lake named Roto-punamu, an extinct spring.

Coming out of the north side of the cave is seen lying picturesquely amongst rocks and bush the spring Rua Kiwi (Kiwi Hole). It is an oblong basin of sixteen feet in length, filled with clear simmering water. The banks of the lake assume here a steep and rocky character; hot springs bubble out of them below the surface of the water, while on the slope are situated, near the Ngawhana spring, the vacated huts of a Maori settlement of the same name, and not far off is the intermittent spring Koingo (the Sighing), the emission of water from which only takes place from three to four times a day, and alternates with the neighbouring Whatapaho.

The above-mentioned springs are the principal ones; on the slope of a hill, rising about 200 feet above the level of the lake, there are more than 100 places that eject steam. South of this steaming hill the banks are lower; on the south-east side of the lake is situated the spring Khakaehu, with which are connected a whole chain of boiling springs, ejecting partly clear and partly muddy water from the swampy ground. In the flats are several small cold-water lakes, and in the back-ground rises a mountain—Te Rangi Pakaru (Broken Heavens)—on the west side of which, from a crater-like hole, there steams a mighty solfatara producing much sulphur.

On the western bank, the great terrace spring—Otuka Puarangi (Cloudy Atmosphere), forms the counterpart of Te Tarata spring. The stalactic steps reach to the lake, and one ascends as on artificially formed marble steps, which are decorated on both sides with green shrubs. These terraces are not so grand as those of Te Tarata, but are more delicate and of a beautiful pink hue, which adds a peculiar charm to this wonderful formation. The basin of this spring is 40 to 50 feet in diameter, and appears as a calm, blue, glimmering, steaming, but not boiling mirror of water. On the northern side, at the foot of the terraces, is the solfatara Whaka-taratara —a sulphur pool in the true sense of the word, from which a hot muddy stream runs into the lake.

There are about twenty-five large hot springs—or *ngawhas*, as the natives call them—at Roto-mahana. I dare not venture to estimate the number of the smaller ones. And Roto-mahana is only one point of a rent above 150 miles long, and 17 wide, between the active crater of Tongariro and that of the White Island in the Bay of Plenty, throughout which hot water and steam are ejected from the earth at innumerable points.

These grand thermal springs have proved most efficient in curing diseases of the skin and rheumatism, so far as the experience of the natives goes; and it is not improbable that in a few years Roto-mahana will be one of the most frequented bathing-places for Australian and Indian invalids. The map is the first that has been compiled of the lakes and springs, and may serve as a guide to the tourist in this interesting district.

Professor Dr. v. Fehling, of Stuttgart, has had the kindness to analyse the waters of the lake and the stalactites. On account of the small quantity of the water, a quantitative analysis could not be made.

A.—Analysis of the Water.

1. Te Tarata Spring, by Mr. Melchior.
2. Rua Kiwi Spring, by Mr. Melchior.
3. Roto Punamu, by Dr. Kielmaier.

In 1000 parts of water was contained:

	1.	2.	3.
Siliceous acid	0,164	0,168	0,231
Chlornatrium	2,504	1,992	1,192
Residue	2,732	2,462	1,726

B.—Silicious Stalactite or Deposit of the different Hot Springs on the banks of the Roto-mahana.

1. Deposit of Te Tarata Spring: (a.) soft, (b.) hard.
2. „ of the Ngahapu Fountain,
3. „ of the Whatapoho Fountain,
4. „ of the Otuka Puarangi Spring.

The analysis executed by Mr. Mayer gives:

	1.		2.	3.	4.
	a.	b.			
Silica	86,03	84,78	79,34	88,02	86,80
Water and organic substance	11,52	12,86	14,50	7,99	11,61
Oxide of Iron } Argillaceous earth }	1,21	1,27	1,34 } 3,87 }	2,99 }	Traces
Chalk	0,45 }		0,27		0,36
Magnesia	0,40 }	1,09	0,26 }	0,64 }	Traces
Alkalies	0,38 }		0,42 }	0,40 }	

MAP V.

WHAINGAROA, AOTEA, AND KAWHIA—THREE HARBOURS ON WEST COAST OF THE PROVINCE OF AUCKLAND.

THE contrast between a weather shore and lee shore coast formation is nowhere so striking as between the west coast and the north-east coast of the North Island of New Zealand. While, from the North Cape to the East Cape, the coast, sheltered from the prevailing winds, presents a most irregular outline, forming deep in the land many indented harbours, navigable by the largest ships — for example, the Bay of Islands and the Waitemata or Auckland harbour — with numerous islands and capes; the West Coast, which is exposed to the westerly wind, is, on the other hand, from Cape Maria Van Diemen to Cape Egmont, an almost regular outline, slightly curved towards the east, and is formed by a nearly uninterrupted chain of sandbanks. These sandbanks in many places, and particularly where there is no steep or higher rocky coast in the background, reach a height of 500 to 600 feet, and when seen from the sea, appear like a chain of mountains. The bays and creeks of the West Coast are, in consequence of these sandbanks, locked up from the sea, and are merely estuaries, navigable only through narrow entrances, in which the sea ebbs and flows. At high water these estuaries appear like large lakes, but at low water immense mud flats, intersected by narrow channels, are laid bare.

On the West Coast are six of these estuaries, three north of the Waikato—the Manukau, Kaipara, and Hokianga harbours; and three south—the Whaingaroa, Aotea, and Kawhia harbours. All these estuaries have this in common—that the sandbanks which are situated before their entrances, are continually shifting their situation and form. This is most prejudicial to navigation, and in consequence all these harbours, with the exception of the Manukau, which alone is navigable by larger vessels, are only available for small coasters.

The most southern of these harbours—Aotea and Kawhia—are represented on this map.

The Whaingaroa harbour is a small sea inlet, six to seven miles long, branching off in many directions, and divided into two parts by a long peninsula. Into the northern bay flows the Whaingaroa river, and into the southern the Waitetuna. The harbour is only navigable for vessels of from 60 to 80 tons, which generally anchor near the outlet; but by boats it is possible to keep up a communication with the most remote branches. At low water the harbour is almost empty; large mud flats are exposed, the narrow channels only retaining water. The Maori population of the neighbourhood amounts to about 400, and that of the European settlers to 122, there being amongst the latter some twenty farmers with their families. About a mile inland from the heads is the township of Raglan. In 1859 it consisted of from six to eight houses, amongst which was, of course, a public-house and a store. Not far from Raglan, also on the south side, is the Wesleyan Mission Station. Opposite, on the north side, is the Maori village Horea, and an old pa.

The borders of the Waitetuna consist of a sandy clay marl, of a tertiary age, containing some, but very few fossils: species of Turritella, Isocardium, and Natica, also a Turbinolia, and some beautiful foraminiferæ. The hills on the south side of the harbour consist of many summits of basalt. Raglan is situated on a soft ferruginous sandstone, which is nothing but hardened sea sand. Opposite to Raglan, on the north side of the harbour, and along the borders, is a most picturesque limestone formation, consisting of tabular masses built up in horizontal strata. Washed and eroded by the sea, these masses assume the most singular shapes: towers sixty to seventy feet high, high walls, columns, &c.

On the south side of the harbour is the Karioi mountain, an extinct volcano of trachydolerite, with a broad and numerously branched summit, which, penetrating far into the sea, forms a very prominent object.

The Aotea harbour is an estuary which, behind its narrow entrance, spreading out into a shallow bay of a width of two to three miles, and a length of six miles, and which, with the exception of a few very small channels, is at low water almost dry. On the west coast is situated the Maori village Rauraukauera, and a Wesleyan Mission School—Beechamdale. Four European families and 270 natives were the whole population in 1859. Dieffenbach reckoned the number of natives, in 1840, at 1200.

The geological conditions are simple and instructive, as the formations seen apart in the Whaingaroa are here placed super-imposed. They can best be observed in a high cliff, situated on the south-east side, and visible from a great distance, called by the Maoris Oratangi, which means that stones fall here with much noise. At the bottom lies a stratum of 40 feet of the same grey clay marl as that of the Whaingaroa harbour, with very few petrifactions. I only found one Inoceramus and a few pectens. Above this marl are large banks of calcareous sandstone, rich in petrifaction. It is the same formation as the tabular limestone of Whaingaroa, the strata varying, some being more sandy, and others more calcareous. At the Puketoa cliff, which stands at the edge of the water, I collected petrifactions belonging to the following genera: Pecten, Spondylus, Cuculaea, Terebratula, Hollicipes, Scalaria, and Schizaster. The marl and sandstone formations make hill land all round the Aotea harbour, which is indented by innumerable small bays. Near the Heads the sandbanks rise to a height of 300 to 400 feet, and traces of lignite may be discovered at high water mark.

The Kawhia harbour is from 6 to 7 miles long, 3 to 4 miles broad, and is intersected by many navigable channels, between which are laid bare at low water shallow mud and sandbanks. The entrance to this harbour is narrowed to only half a mile by a far extending land tongue—Te Maika. At the entrance are bars, which confine the navigation of the harbour only to smaller craft. The coasting trade is partly carried on by Europeans and by Maoris. Six European families are settled on different localities of the harbour, and the number of natives were in the year 1859 from five to six thousand.

The steep and abrupt coast wall of the south side of the harbour, in the neighbourhood of Takatahi, is built up of steep strata of sandstone and calcareous marl. It was here I had the pleasure of finding the first New Zealand ammonites and other petrifactions. (Ammonites now Ocelandicus, Inoceramus Haastii, &c.) In the south-westerly direction of Takatahi, also on the south side, is Ahuahu, a land-point on the Waiharakeke channel, in the neighbourhood of the Wesleyan Mission Station, where there is a rich mine of belemnites (Belemnites Aucklandicus var. minor). The cliffs are of clay, of a greenish brown colour, the steep strata of which alternate with hardened lime marl. At low water it is possible to collect the belemnites at the foot of the cliffs in great numbers. The natives call them Roke-kanae, which means the excrements of the fish kanae.

The whole southern borders of the Kawhia Harbour consists of strata containing belemnites and ammonites belonging to the Jurasic system. The same tertiary argillaceous marl and limestone which appear at the Aotea and Whaingaroa Harbours appear also at the northern bank of the Waiharakeke River, forming the border walls and distributed in almost horizontal strata over the whole south-eastern side to the Awaroa River. At the Rakaunui River these chalk banks reach to the water's edge, and form along the coast the most picturesque rocks in the shape of towers and ruins, in consequence of which this part of the Kawhia Harbour has been designated the New Zealand Switzerland. The romantic and various shapes assumed by the torn and worn masses of rocks surprises the eye everywhere, while in the valleys, where lie the villages of the natives, the rich fields of corn and maize delight the beholder. I consider this as one of the most beautiful and fertile districts of New Zealand which I have seen. The character of the landscape remains the same far up into the mountains, and 1000 feet above the harbour white masses of rock penetrate through the verdure of the forest and bush. Hence the name of Castle Hills for these mountains, which the natives call Whenuapu. This neighbourhood also possesses numerous caves. This limestone formation appears also on the north side of the Kawhia Harbour, in the Towara Bay, and on the Puti River; it is rich in large oysters and terebratulæ.

The east side of the harbour consists partly of scattered volcanic tufa and conglomerate, which are connected with the trachy-dolerite chain of Pirongia. Over these mountains, paths thickly covered with bush lead inwards to the Waipa valley.

III.

LECTURE ON THE GEOLOGY OF THE PROVINCE OF NELSON.

BY DR. F. VON HOCHSTETTER.

LADIES AND GENTLEMEN,—It is with much pleasure that I respond to the wish expressed by you, and at the same time fulfil my promise of communicating the results of my geological explorations in a lecture on the geology of this Province; and it is with a feeling of pride that I see so large and distinguished an assemblage met here this evening.

On my arrival in Nelson, in the beginning of the month of August, I hardly hoped to be able to extend my researches so far as to obtain an accurate idea of the geological features of the Province. The time allotted to me was very short; the geological field of the Middle Island, on which I was entering, was, in comparison with that of the Northern Island, an entirely new one. Entering into Blind Bay upon a bright morning, I saw all round me lofty snow-covered mountain chains. It was the middle of winter, and I doubted whether at this season of the year extended geological researches were possible. This doubt was soon removed; the glorious weather which favoured my excursions gave me full confidence in the far-famed and deservedly-praised Nelson climate. My first exploration opened up to me a field at once so interesting as regards scientific research, and at the same time of so great practical importance, from the existence of those very valuable substances, gold, coal, and copper, that, in order to give greater value to the results of my observations, I willingly resolved to respond to the wish of the inhabitants, and to remain a month longer among you.

I feel myself in the highest degree obliged to the inhabitants of this Province, who, so soon as the 'Novara' arrived in Auckland, invited the members of the expedition to visit Nelson, for the honourable and hospitable reception, and for the active assistance in the prosecution of my objects, which I have met with on all sides; and I wish to take this opportunity of expressing my thanks to the Provincial Government for the admirable arrangements which on its part were made so as to extend to the utmost limit the sphere of my explorations, and enable me to occupy to the greatest advantage the limited time at my disposal.

Allow me, before proceeding further, to give you an account of my different journeys, and to detail to you the places which I have visited.

I began in the immediate neighbourhood of the town of Nelson, by a short excursion to Brook Street Valley, and a visit to Mr. Jenkins' brown-coal mine. I then proceeded in the 'Tasmanian Maid,' which the Government had chartered for this extra trip, to Croixelles Harbour and Current Basin, and examined the veins of copper ore which show themselves there.

We proceeded up Current Basin as far as the French Pass, and on our return landed in the Bight of Owhaua, on the south-eastern corner of D'Urville's Island, where copper ore is also found. From thence we steamed, without loss of time, during the night, across to Golden Bay, where I went on shore at Collingwood, and visited the gold-fields and the bone caves of the Aorere valley. Thence I proceeded along the coast to Pakawau, and examined the coal-field there, and the graphite which is found in the hill at Taumatea. Returning overland from Golden Bay to Nelson, I visited, on the way, the Parapara gold-field, the brown-coal deposit at Motupipi, followed the course of the Takaka Valley upwards, crossed the mountain range that divides the Takaka and Riwaka valleys, and passing through Motueka, reached Nelson by the Moutere and the Waimea. Another day was devoted to an examination of the Boulder Bank and the Arrow Rock. I next proceeded by the valley of the Maitai to an examination of the Dun Mountain. I next visited the Wakapuaka district and the Happy Valley; and, at a later date, in an opposite direction, spent some time in examining the fossiliferous schists of Richmond and the Wairoa Valley.

After I had made myself acquainted with the geological relations of the nearer lying districts of Golden and Blind Bays, arrangements were made for a more distant excursion in a southerly direction, to the Wangapeka and the Lake country, and in an easterly direction towards the Pelorus, the Wairau, and the Awatere valley. My time was too limited to enable me personally to undertake both these geological explorations. I therefore availed myself of the friendly co-operation of my friend and companion, Mr. J. Haast, who has hitherto accompanied me on all my journeys in New Zealand. My friend Haast proceeded by the 'Tasmanian Maid' to Queen Charlotte Sound, landed in Maraetae bight, examined the coast as far as Waikawa, and proceeded overland to Waitohi: thence by the Waitohi Pass along the Tua Marina to the Wairau plain. Thence by the Taylor Pass Mr. Haast proceeded to the Awatere, returning by Maxwell's Pass to the Wairau. After an examination of the Waihopai valley, he proceeded through the Kaituna to the Pelorus, and returned by the Pelorus road to Nelson. I am indebted in the highest degree to my friend Haast for the interesting and important information which he has communicated to me concerning this region, and for the disinterested zeal and the ability with which he carried out his task of contributing to a knowledge of the geological relations of the country visited, and also for a valuable addition to my collections.

I myself took my way in a southerly direction towards the Motueka and Wangapeka valleys; crossed the chain of hills to the Buller River; followed this upwards to the Rotoiti Lake; from thence made my way to the Top-house, in the Wairau valley, and returned by the Big Bush to Nelson.

I am thus enabled to say, that it has been possible for me to obtain a general geological view over the whole of the northern half of the Province of Nelson, from the Awatere valley on the east to the Aorere valley on the west; while the cross valley of the Buller River, between the Rotoiti Lake and the gorge of the western mountain chain, indicates the southerly limit of the district explored.

I have much to thank my Nelson friends for, both in the way of information and contribution to my collections, and am at the same time indebted to the various gentlemen who, in a spirit of friendship, accompanied me on my journeys.

My best thanks are also due to the various settlers in whose houses I have found such hospitable quarters. May I be allowed, without mentioning individual names, to express my most sincere thanks to all these gentlemen for their active assistance and valuable contributions to the 'Novara' collection.

I have begun to put together on a map the results of my observations, with the view of laying the foundation of a Geological Map of the Province of Nelson. So soon as time will allow me to complete this map, 1 will hand over to you a copy of it with pleasure; at the same time expressing a wish that the numerous friends of geology among you, and if they will allow me to say it, my geological scholars here, may continue it and improve it, where I, either from want of time or inaccessibility of the district, have not been able to fill in the details.

I will now come to the subject matter of my lecture.

I.—PHYSICAL FEATURES.

The character of the surface is always more or less indicative of the geological structure of a country. Even to those who have not deeply studied the science, the different forms which mountain ranges show, will indicate a different geological formation. The difference in these external appearances of the country is very striking, if you come from the Northern to the Southern Island.

In contrast with the comparatively low plateaus extending over the largest part of the Northern Island, and broken only by high volcanic peaks, you find on the Middle Island lofty and abrupt mountain ranges, striking in long parallel chains, divided by deep longitudinal valleys, and broken at right angles by rocky gorges. This complication of rock and gorge runs as the great backbone of the island from north-north-east to south-south-west, and from strait to strait. Well do you name it your "Southern Alps." Amongst them rises in grandeur a mountain named after the great discoverer of the South Sea, Mount Cook, of a height equal to Mont Blanc. It towers above the rest, crowned with perpetual snow, with ravines glistening with glacier ice. To the steep perpendicular cliffs with which the Southern Alps breast the stormy sea on the west coast, are opposed fertile plains extending along the eastern shore.

From a central point, which (near the boundary line of the two Provinces of Canterbury and Nelson) gives rise to the Hurunui and Waiau-ua Rivers, flowing to the eastward, and to the Grey and Enungahua, flowing to the westward, the Southern Alps send forth two arms through the Province of Nelson, the extremities of which are washed by the waters of Cook's Straits. These arms are again subdivided by longitudinal valleys into numerous ranges, with peaks from five to six thousand feet high. I will distinguish between the two arms by giving the name of the "Western Ranges" to those which, with a northerly strike, terminate in Massacre Bay, between Separation Point and 'Cape Farewell, and the name of the "Eastern Ranges" to those which, running in a north-easterly direction, terminate in the Pelorous and Queen Charlotte Sound.

In the acute angle between the two ranges are situated the Lakes Rotoiti and Rotorua, from which undulating hills, intersected by numerous streams, gradually slope from an altitude of 2,000 feet to the plains of the Waimea and the shores of Blind Bay.

I can hardly remember a more beautiful and more striking scene than when I first looked, on a clear winter day, from a high point on the Richmond Hills over the fertile Waimea plains, lying like a map beneath my feet, studded with homesteads and covered with cultivations, towards that triangle of snow-capped ranges.

It is, without doubt, in consequence of the peculiar configuration of the mountain ranges, that Blind Bay is favoured with an extraordinarily temperate climate. The western and the eastern ranges of Nelson, converging towards the south, form a regular wedge, which diverts on the one side the force of the south-westerly winds, and on the other side the force of the south-easterly winds. Those parts of the Province of Nelson which are not enclosed between the legs of the triangle, do not enjoy the same serenity of climate. In Golden Bay and in the Wairau country, which lie respectively to the west and to the east, in the line of the bounding ranges, gales of wind and bad weather generally are much more frequent than in Blind Bay.

The "spout wind," blowing with considerable violence

during the summer from the south, is a local wind of Blind Bay, due to the same physical configuration of the country. The calm heated air of the Waimea plains and of the low hills, rising in obedience to physical laws into the higher levels of the atmosphere, is suddenly replaced by volumes of colder and denser air, which rush down towards the plains from the mountain ranges behind.

I have made these remarks in order to offer an explanation of some of the most striking peculiarities of the Nelson climate —the Montpellier of New Zealand.

II.—GEOLOGICAL FEATURES.

The western and the eastern ranges of Nelson are totally different in their geological character. The western ranges consist of primitive formation, being built up of old crystalline schists, or metamorphic rocks. The eastern ranges are the oldest sedimentary strata, primary formation, broken through in places by masses of plutonic rocks. The lower undulating hills lying in the angle between the two ranges are nothing but an immense accumulation of debris from the mountain ranges on either side, rolled together by the action of the sea, which in former ages washed the bases of the mountains.

When I say *Gold* in the western ranges, *Copper* in the eastern ranges, and *Coal* in the basins between them, I have indicated the chief mineral characteristics of the region referred to. I will now speak more in detail of the

(1) *Primitive Formation of the Western Ranges.*

Taking a cross section from east to west, through the western ranges, we find the sub-divisions of the primitive formation succeeding to one another in their normal geological order.

(*a.*) *Gneiss and Granite Zone.*—The western shores of Blind Bay, from Separation Point to Riwaka, consist of granite, bordered on the eastern side opposite to the Tata Islands by gneiss. This same zone of granite and gneiss may be traced in a southerly direction up the Motueka River to the confluence of the Wangapeka, and is cut through by the Buller River, where it enters the gorge of the Devil's Grip on the western ranges, and extends all along the eastern slope of the mountains as far as the Rotorua Lake.

(b.) *Zone of Hornblende Schist and Chrystalline Limestone* (*Urkalk*).—Proceeding from the granite and gneiss towards the west, we next meet, on the top of the Pikikerunga range, between Riwaka and Takaka, a broad zone, on which hornblende-schists, quartz-schists, and crystalline limestone succeed one another in regular and numerous alternating strata, with a vertical dip and a strike nearly due north and south. This formation continues in a westerly direction to the opposite side of the Takaka valley, where it is broken through by eruptive masses of diorite-porphyry and serpentine, which show themselves in the Stony Creek and Waingaroa. The same zone of crystalline schists exhibits itself in the steep escarpments of the gorge of the Wangapeka. A characteristic feature of this limestone formation is the existence of numerous funnel-shaped pits, which have been hollowed out by the action of water, which has dissolved the limestone.

The interesting phenomenon of the Waikaromumu springs in the Takaka valley, where whole rivers suddenly appear on the surface with the water bubbling, is readily explained by a subterranean passage of the water through the limestone from the ranges. This crystalline limestone on the ranges must not be confounded with the other limestone in the Takaka valley, which belongs to the tertiary period.

(c.) *Mica Schist and Quartz-Schist.*—The crest of the western ranges, with peaks rising to an altitude of about 6,000 feet, the Anatoki mountains, Mount Arthur, and the chain lying between the source of the Wangapeka River and the Buller river gorge, consists of mica schists, containing garnets, alternating with quartz schists.

(d.) *Zone of Clay Slate.*—Still proceeding towards the west, the mica-schists pass, by insensible gradations, into clay slates, which, however, still exhibit the same alternating strata of quartz-schist. The Aorere valley and the lofty peaks on its eastern side, as the Slate River peak, Lead Hill, Mount Olympus, and the Haupiri range, generally belong to the clay-slate zone. In all these ranges the strata are more or less vertical, and exhibit unmistakable signs of great disturbance at former geological periods. For instance, Mount Olympus presents the peculiar appearance of strata diverging from below towards the serrated edge of the mountain, like the folds of a fan. A similar disposition of strata is observed on the loftiest summit of Europe, namely, on Mont Blanc.

Gold.

In the mica-slate and clay-slate zone of the western ranges, we have the matrix of the gold. From the interest attaching to this subject, I may be allowed to repeat the limits of these gold-bearing formations. On the east these formations are bounded by the Takaka valley; on the west by the Aorere valley, so that its breadth is from fifteen to twenty miles, and includes the Anatoki and Haupiri ranges. In a southerly direction the same formations can be traced to the gorge of the Buller river. How much further it extends in that direction has not yet been ascertained; but inasmuch as gold has been found at the northern extremity of the Southern Alps, and also in the gravels of the Mataura, in the Province of Otago, towards the southern extremity of the backbone, it is not unreasonable to infer that the same gold-bearing zone may extend continuously throughout the whole length of the Middle Island.

Before speaking more specially of the gold-fields, I wish to correct some of the theories popularly current among the diggers, according to which gold is to be traced to the action of fire. The gold, in its original position, is in larger or smaller particles dispersed throughout the quartzose constituents of the mica and clay-slate formations. By the gradual wearing away of these rocks, through the action of the elements, extending over immense periods of time, large masses of debris have been formed, and nature itself has executed an operation of gold-washing, by collecting the heavier particles and depositing them in the gullies of the streams, or in the conglomerates covering the slopes of the hills.

There are, therefore, two principal descriptions of diggings; either "river diggings," in the beds of the streams, or "dry diggings," in the conglomerate and gravel accumulated on the slopes of the mountains.

I will first describe the best-known and most worked of your gold-fields—namely, the Aorere and the Parapara gold-field.

The Aorere and Parapara Gold Field.—You are all aware, that the gold in the Aorere valley is confined to the eastern side of the valley; the only traces of gold found on the western side are on the Kaituna stream, but not indicating any rich deposit on that side, which, as fertile agricultural land, must be left to the farmer. You know that all the tributaries of the Aorere river proceeding from the Haupiri range, as, for instance, Appoo's River, the Slate River with its different branches, the Boulder Rivers, Salisbury Creek, and also the Parapara River, which proceed northwards from the same range, have been more or less successfully worked by various parties of diggers. The rounded nature of the gold particles shows that the gold has been brought down by water; and the fact that the heaviest gold is found in the upper parts of the streams, points clearly to the mountains as the source of the metal.

But it would be improper to speak about an Aorere goldfield, if the gold were confined to the deep and narrow gorges of the streams, cut down into the clay-slate rocks.

The whole region of the eastern side of the Aorere valley, rising from the river bed towards the steep sides of the mountains at an inclination of about eight degrees, and occupying from the Clarke River towards the south, to the Parapara on the north, a superficial extent of about 40 English miles, is a gold-field. Throughout this whole district, on the foot of the range, we find a conglomerate deposited on the top of the slate rocks, reaching in some places a thickness of twenty feet. Pieces of driftwood changed into brown-coal indicate a probably tertiary age of this conglomerate formation. Where a ferruginous cement binds the boulders and the gravel together, this conglomerate is compact; in other places only fine sand lies between the larger stones. Quartz and clay-slate boulders are the most commonly met with. This conglomerate formation is not only cut through by the deep gullies of the larger streams, but in some places washed by the more superficial action of occasional water, and so divided into parallel and rounded ridges, of which that portion of the district called the Quartz Ranges is a characteristic example. This conglomerate formation must be regarded as the real gold-field, prepared in a gigantic manner by the hand of nature, from the detritus of the mountains, for the more detailed and minute operations of man.

While the less extensive, but generally richer river-diggings afford better prospect of gain to the individual digger, the dry diggings in the conglomerate will afford remunerative returns to associations of individuals who will work with a combination of labour and capital. The intelligent and energetic gold-digger, Mr. Washbourn, is the first person who has proved the value of the dry diggings in the Quartz Ranges, and has demonstrated the fact that gold exists in remunerative quantities in the conglomerate. I am indebted to Mr. Washbourn for the following interesting details. He writes to me as follows :—" In the drives into the conglomerate of the quartz ranges, the average thickness of dirt washed is about two feet from the base rock ; and the gold produced from one cubic yard of such earth would be, as nearly as I can calculate, worth from twenty-five to thirty shillings. This includes large boulders; so that a cubic yard of earth, as it goes through the sluice, is of course worth more, as the boulders form a large proportion of the whole. Where the earth is washed from the surface to the rock, the value per cubic yard is much less—not worth more, perhaps, than from three shillings to six shillings per yard, and it would generally pay very well at that."

With this data, the following calculation may be made. We will reckon the superficial extent of the Aorere and Parapara gold-fields at thirty English square miles ; the average thickness of the gold-bearing conglomerate, at a very low rate, at one yard ; and the value of gold in one cubic yard at five shillings. Upon this data, the value of the Aorere gold-field is £22,500,000, or £750,000 for one square mile.

I am not a practical gold-digger myself, but I will leave it to those who are more versed in that pursuit to contrive the means by which this wealth may be best extracted from the soil. Considering that Mr. Washbourn was able to pay his men wages from ten to twelve shillings a day, and still to make a considerable profit, the richness of the deposit of gold in the conglomerate is clearly proved.

You may allow me to add, from inquiries I made on the spot, the number of diggers working on the Aorere and Parapara diggings is not more than about two hundred and fifty. Although the diggers cannot be at work continually, a large portion of time being occupied in bringing their provisions across a rugged country ill-provided with roads, and occasionally stopped by floods in the rivers, it is considered that a digger earns on an average twelve shillings a day.

The history of the gold-field does not record any large fortunes made by single diggers, but steady average gains. The largest nugget found was in the Rocky River, a nugget of 9 oz. 18 dwts.

The whole produce of the gold-field from the beginning, in 1857, up to the middle of August, is recorded, in the General Government Gazette, as about £150,000.

I may add that, looking to the position of the gold-field generally, and its proximity to the sea, there is probably no other gold-field which, with moderate outlay upon roads, could be made more easily accessible, or might afford greater facilities for being worked. I have very little hope that quartz reefs will be found in this district rich enough to pay for crushing.

The country on the western side of the gold-bearing ranges, further south than the Clarke River, has not yet been perfectly explored with regard to probable gold-fields, and I proceed, therefore, to the eastern side of the gold-bearing formations.

I may remark that there is no foundation for the belief, so generally entertained amongst diggers, that gold-fields are only found on the western side of ranges, and not on the eastern.

The Anatoki and Takaka Diggings.—From the same mica-slate and clay-slate zone, from which on the western side the gold-bearing branches of the Aorere valley run, on the eastern side the Takaka river with its branches takes its rise. It is therefore not surprising that gold is also found on those rivers. If the farmer settled on the rich alluvial plains of the Takaka finds markets bad, he has but to ascend to the higher parts and branches of the river to fill his pocket. Gold is found in sufficient quantity to pay river diggings in the upper Anatoki, Waingaro, and Takaka, the heaviest nuggets in the Waitui river, which takes its rise from the Mount Arthur range. In the Anatoki valley a quartz reef is spoken of which promises well. The interesting metal, *osmiridium*, as has been proved by specimens forwarded for analysis to Mr. Clarke, of Melbourne, is a peculiar accompaniment of the Takaka gold,

Titaniferous iron, magnetic iron, and garnets—not rubies as generally thought—are everywhere found on the river diggings of the Province. It must be left to the energy of future explorers to determine if there be not, as it is most probable there is, a similar gold-field as the Aorere gold-field, hidden under the dense forests on the eastern slope of the ranges.

Wangapeka.—With a view to exploring the country lying to the south of the Takaka, on the eastern side of the gold-bearing formations, I made a journey to the Wangapeka. My guide to that country, most difficult of access, was Mr. Clarke, who had formerly been prospecting there for gold. On this occasion I had the pleasure of the company of the Superintendent. The Wangapeka, as large if not larger than the Motueka, near its junction with the Sherry River, runs through a wide terraced valley.

The hills bordering the valley are composed of tertiary strata, marl, sandstone, and limestone. At places on the sides of the valley, granitic rocks show themselves as the foundation of the tertiary strata. The boulders and shingle brought by the river from the deep gorge, through which it enters the broad valley, prove, on examination, that the river takes its origin in a zone of hornblende-schists, and crystalline limestone, the continuation of the formations between Takaka and Riwaka. There is therefore no reason to expect an auriferous river-bed. I might here mention that this valley seems the peculiar home of wild pigs, the immense number of which have rooted up the whole surface. The wet weather we experienced prevented my exploring those rivers which take their rise further westward, in the mica-slate and clay-slate ranges as I expect, as, for instance, the Batten River and its branches. It was here that Mr. Clarke found the best result of his prospecting expedition. He found not only gold, but, on the edges of the tertiary formation towards the crystalline ranges, large seams of coal cropping out.

As a very probable gold country, I should recommend the exploring of the high range situated between the sources of the Wangapeka and the gorge of the Buller. That range is, so far as I can judge, the continuation of the Mount Arthur, Anatoki, and Haupiri ranges.

I shall hereafter find an opportunity to remark upon the

Motueka diggings, and will conclude this portion of my lecture by stating that the Nelson gold-fields are a fact, and that which is at present known is but the beginning of a series of discoveries which time will bring to light.

With regard to other minerals in the western ranges, there are no indications of quicksilver, as it was supposed, But Mr. Skeet informed me that pieces of lead ore are found in the Waingaro River; and large masses of brown iron ore, which have been mistaken, from their somewhat similar appearance, for scoria, are deposited at the Parapara. This has given rise to the idea of the Parapara being volcanic.

(2.) *Primary Formations in the Eastern Ranges.*

The eastern ranges are of an entirely different geological formation to those just described in the west; old primary slates and sandstones, of very various character, form lofty ridges, intersected by parallel longitudinal valleys. The strata are all more or less vertical, and the parallelism of their strike from north-east to south-west continues with remarkable regularity. One and the same stratum can be traced from Cook's Straits to the far interior in the south.

In the central ridge, which has its northern termination in Mount Stokes, between the waters of the Pelorus and Queen Charlotte Sound, the slates exhibit a more crystalline character. At Ship's Cove and Shakespeare Bay in Queen Charlotte Sound, in the Kaituna Pass and other places, almost crystalline micaceous clay-slates, with quartz layers and veins, occur.

On either side of this central ridge the slates exhibit a more sedimentary character, alternating with dioritic-schists, with amygdaloids, with very compact sandstones, approaching the character of graywacke. As no fossils have yet been found in those oldest sedimentary New Zealand schists, it is impossible to assign to them their geological place in a European classification of strata.

The slate and sandstone ridges are flanked by serpentine.

Below the confluence of the Blarich River with the Awatere, where the side of the mountain has slipped with an earthquake rent, serpentine appears. The Grey Mare's Tail is a waterfall over a serpentine cliff. The serpentine extends, in a south-westerly direction, through the Blarich valley towards Mount

Mowatt, whose south-eastern slope to a height of about 2000 feet is composed of serpentine. In the bed of the Blarich River, Mr Haast found a piece of copper ore of the same description as the Dun Mountain ores.

On the western side, the serpentine occurs developed to a much greater extent. An immense serpentine dyke, of a thickness of several miles, stretches from the northern extremity of D'Urville's Island, across the French Pass, through the Croixelles, by the Dun Mountain and Upper Wairoa, and is met with again, on a continuation of the same straight line, on the Red Hills, near the Top-house, on the northern side of the Wairau valley. This dyke can thus be traced from north-east to south-west for a distance of eighty miles. The strike of the serpentine dyke is perfectly parallel to that of the slates, but its eruptive origin is proved by the occurrence of a breccia of friction (Reibungs breccia) at the line of contact, and the fact of beds of slate enclosed in it being converted into hard and semi-vetrified cherts. The serpentine, in its turn, has been broken through by eruptive dykes of hypersthenite and gabbro. The rock of the Dun Mountain proper is a variety of serpentine, of so novel and peculiar a character that I am obliged to apply to it a new term, and call it "Dunnite." The Dun Mountain district offers to the scientific geologist a field of unbounded interest; but I shall perhaps best respond to the wishes of my audience by telling them something about the ores of copper and chromate of iron which are the characteristic metals of that serpentine dyke.

Copper.

The occurrence of native copper, red oxide of copper, and copper pyrites, the principal copper ores of the Dun Mountain, is by no means peculiar to the serpentine of New Zealand. In the serpentine district of Cornwall, for instance, native copper is found. The Monte Ramazzo, near Genoa, contains copper ores in serpentine, and in North America the same thing occurs.

I have visited (accompanied by Mr. Hacket and Mr. Wrey) all the workings of the Dun Mountain. I could not convince myself of the existence of a number of parallel lodes, so as to justify the various names which have been given, and which appear to designate different lodes. The Dun Mountain copper ore does not occur in a regular lode ; by which I mean a metalliferous dyke of different mineral composition from that of the rock of the mountain. As is usual in serpentine, the copper ore occurs only in nests and bunches. The richer deposits of copper ore form lenticular shaped masses, which, when followed, may increase to a certain distance, but then disappear again in a thin wedge. Where these nests are large and rich, one alone may sometimes make the fortune of a mine. The richest found on the Dun Mountain appears to have been that of the Windtrap Gully, from which pieces of native copper (some of them weighing as much as eight pounds) were extracted. These nests of copper ore occur in the Dun Mountain in one continuous line, as if a rent had taken place in the serpentine rock, into which copper had either been injected from beneath, or deposited there by the operation of some causes which science is unable to explain. The green and blue silicates of copper are surface minerals, which are only of value by showing the direction of the fissure in which the real ore may be looked for at a greater depth. At a certain distance below the surface, they disappear entirely, and it is only by the broken and softened character of the serpentine that the miner is enabled to follow the fissure from one deposit of metal to the other. The occurrence of the best indications of copper ore on the surface over a continuous line of about two miles, affords good ground for supposing that considerable quantities of ore are contained in the mountain ; but, on the other hand, owing to the manner in which the ores occur in isolated bunches, mining operations in such a region are always attended by less certain profits than where the metal is deposited is a regular lode ; and I may be allowed to express a hope that the Dun Mountain may prove to be all that the Nelson people could wish.

In Croixelles and in Current Basin, where copper mining operations have been carried on, the indications were very obscure, and the result has proved that there is no reasonable ground to expect a profitable copper mine there. More promising specimens of copper ore have been obtained from D'Urville's Island. The character of the ores met with there is quite the same as in the Dun Mountain.

I will add a few words about chromate of iron. This

mineral is an ordinary accompaniment of serpentine rock, and occurs in the Dun Mountain in great force. Of its commercial value I do not feel myself qualified to speak, but should its value be considerable, the abundance of it is so great that it must prove a source of much wealth to the mine.

Having described the central parts of the western ranges, and the serpentine which flanks it, there still remains to me to describe a zone of old sedimentary rock, which lies between the serpentine on the east, and Blind Bay and the Waimea plain on the west. The best section of this zone is obtained by following up the course of the Maitai to the Dun Mountain. Immediately to the west of the serpentine we meet with a belt of calcareous schists, which attains its highest elevation on the summit known as the Wooded Peak, and continues on its strike parallel with the serpentine dyke. Proceeding to the westward, we pass over a band of greenish and reddish coloured slates, of a thickness of about five English miles. The same description of slates continues all along the ranges, as far as the Big Bush road to the Wairau. The absence of any fossil remains in the calcareous schists and in the slates prevents me from assigning to them with confidence their geological age. I give them therefore a local name, and call them the Green and Red Maitai Slates. In places these slates are broken through and altered by eruptive rocks, as, for instance, in Brook Street valley by diabase, and near Wakapuaka by sienite.

(3.) *Secondary Formations.*

Between Nelson and Wakapuaka, black slates and shales are found close to the edge of the water. In these we find the first indications of organic remains. Of the nature of these organic remains I have not been able perfectly to satisfy myself; they appear, however, to belong to the vegetable kingdom, and have more resemblance to sea-weeds than anything else.

In the same line, further south, the Richmond sandstones form the boundary of the western ranges. No less interest attaches to these sandstones, which contain many and perfect fossil molluscs, and are, so far as I know, the oldest fossiliferous strata in the Province. The fossils belong to the genera Mytilus, Monotis, Avicula, Spirifer, Terebratula, which seem to indicate a secondary age for the formations. If I were to trace any analogy between these strata and any European formation, I should say that they occupied in New Zealand the place filled by the Muschelkalk in Europe.

I have described now the formations of the higher ranges of the Province. Before leaving them I will observe that they possess an extraordinary interest for the botanist. Dr. Monro and Dr. Sinclair have brought from those regions specimens of the greatest interest, and new to science. And a large field is still open for those who will follow in their steps.

Zoologists may be surprised to hear that on the top of limestone ranges between 3,000 to 4,000 feet high, at the Pikikerunga and the Maunga-tapu, a large land snail, or helix, is found, as large as the Helix Busbyi of the Northern Island. Mr. Skeet found a live specimen on the Anatoki mountains; and to Mr. W. Askew, at Riwaka, I am indebted for a perfect specimen of that new and rare shell.

Pukawau Coal Fields.

I come now to speak about the Pakawau coal-field, as probably belonging to the secondary period. The Pakawau coal-field overlies the mica and clay-slate formations of the western ranges. The Pakawau stream exposes various strata of the coal-field, its conglomerate, sandstone, shales, and seams of coal. There have been workings on the exposed seams on both sides of the stream. A quantity of coal extracted from a seam of four feet thickness on the north side, which has lain exposed to the weather for two years, and still remains in the condition in which it was extracted, at once convinced me of the difference existing between this coal and the other New Zealand coals which I have seen. The coal is a dense, heavy, black coal, of a laminated structure, breaking in large pieces which do not crumble. In the evening I burned the coal in a fireplace, and was pleased with the large amount of flame and heat given out by it, without sulphureous or other disagreeable smell. It burned away to a clean white ash. Mr Curtis has kindly forwarded to me an analysis of this coal, made in the year 1853, by Mr. Theoph. Heale, at Auckland. Mr. Heale proved the excellent qualities of the coal as a gas coal; the quantity of carbon (not more than 53 per cent.) would not confer upon this coal a high character as fuel; but this low per-centage probably arose from the piece submitted to analysis being mixed with shale. To me it appears that the coal must contain at least 70 per cent. of carbon, and that it will be found a very excellent coal for steam purposes.

On the southern side of the stream, the old workings exhibited the following sections:—

Shale					
Coal	...	...	...	0ft.	5in.
Shale	...	...	...	0	$3\frac{1}{2}$
Coal	...	...	...	0	$4\frac{1}{2}$
Sandstone	...	...	0	2
Coal	...	...	...	1	2
Shale					

Coal, in all, 2 feet.

Thus, the natural sections and the old workings show various seams, but none of them of great thickness, and in all of them more or less bands of shale.

The dip of the seam is towards south-west—that is, towards the West Wanganui harbour, at an angle of twenty degrees, and the coal-field reaches, undoubtedly, from Pakawau to West Wanganui. In a coal-field of such extent, it may be with confidence affirmed that seams of much greater thickness exist, and the way to ascertain their existence, is to make borings. That is the first thing for any company to do which undertakes to work this very valuable coal-field. My reason for assigning to this coal-field a secondary age, is the existence of impressions of fossil plants, referable to calamites, ferns, and dicotelydones. Although the Pakawau coal-field does not belong to the carboniferous period, experience will show that the coal will rank in quality with the black coals of older date.

I proceed from these older coals to the tertiary period and the brown-coal formation.

(4.) *Tertiary Formations.*

The tertiary formations which I observed in the districts of Golden and Blind Bays belong to that group which I mentioned in my Auckland lecture as the older one. All the wide valleys and basins which from the shores of Cook's Straits run inland between the high primitive and primary ranges, are filled with tertiary strata, which at some places attain an altitude of 2000 feet.

This formation is divided into two parts: the lower, or a brown-coal formation; the upper, fossiliferous marl, sandstone, and limestone.

I will give a short description of these strata, from Cape Farewell to Awatere.

It is a remarkable fact that at Cape Farewell, the north-westernmost point of the Middle Island, where the sea swarms with echinides, commonly called sea-eggs, the tertiary sandstone cliffs are also found full of fossil remains of the same family, but differing in species.

In the Aorere valley the original tertiary strata are, by later action, for the most part destroyed. On the western side of the valley indications of brown-coal have been found. On the cliffs of Collingwood, marls, containing few fossils, are the representatives of the formation.

Higher up the valley large isolated masses of a fossiliferous calcareous sandstone, or, if you will, of a sandy limestone, penetrated by numerous caves, are the remains of a once continuous tertiary stratum.

The caves above Washbourn's Flat, in these isolated limestone blocks, have lately become famed as bone caves, the cemeteries of gigantic birds, which, in the traditions of the Maoris, are remembered as the frightful Moas, and which to science are known as the genera of *Dinornis, Notornis,* and *Palapteryx.*

When, in 1857, I saw in the British Museum the skeletons of *Dinornis elephantopus* and *Dinornis robustus,* I little thought that I should so soon be in possession of the same treasures.

Before my arrival at Collingwood I had heard of the late discovery of Moa bones in those caves, and I was anxious to procure those specimens, which I had had so little success in obtaining in the Northern Island.

In the first cave which I entered—my friend Haast has since given it my name—after a short search, I dug out fragments of bones from the loam on the bottom of the cave. I was convinced that the treasures had not all been carried away, as from the caves in the Northern Island; and on the same day the finding of a Moa skull—so far as I know the most perfect yet found in New Zealand—was the reward of further researches.

Being obliged myself to leave for the Pakawau coal-field, my friend Haast remained behind in company with the young surveyor, Mr. Maling, to make more extensive researches. The bottom of a second cave, the Stafford's cave, was turned up, and the bottom of a third one, the Moa Cave. The excitement of the moa-diggers was great, and increased; for the deeper they went below the stalagmite crusts covering the floor, the larger were the bones they found, and whole legs, from the hip-bone to the claws of the toes, were exposed. They dug and washed three days and three nights, and on the fourth day they returned in triumph to Collingwood, followed by two pack-bullocks loaded with moa bones.

I must confess that not only was it a cause of great excitement to the people of Collingwood, but also to myself, as the gigantic bones were laid before our view. A Maori bringing me two living kiwis from Rocky River gave us an opportunity to compare the remains of the extinct species of the family with the living *Apterix*.

It gives me much pleasure to acknowledge the zeal and exertions of my countryman and friend Haast, in adding such valuable specimens to the collections of the 'Novara' expedition. The observations of M. Haast, made during this search, throw a new light upon this great family of extinct birds. He found that according to the depth so was the size of the remains, thus proving that the greater the antiquity the larger the species. The bones of *Dinornis crassus* and *Palapterix ingens* (a bird standing the height of nine feet) were always found at a lower level than the bones of *Dinornis didiformis* (Owen) of only four feet high.

I have the pleasure of showing you here a leg of *Dinornis crassus*.*

I have since had my collection of bones increased by various

		Long.	Circumference of the Shaft.
* Tarsus		9¼ in.	6·9 in.
Tibia		22	6·6
Femur		13	8·0
Spread of the claws		15	

contributions from Messrs. Wells, Haycock, and Ogg, and a nearly perfect skeleton of *Palapterix ingens* presented by the Nelson Museum to the Imperial Geological Institution of Vienna.

These gigantic birds belong to an era prior to the human race, to a post-tertiary period. And it is a remarkably incomprehensible fact of the creation, that whilst at the very same period in the old world, elephants, rhinoceroses, hippopotami in South America, gigantic sloths and armadillos; in Australia, gigantic kangaroos, wombats, and dasyures were living; the colossal forms of animal life were represented in New Zealand by gigantic birds, who walked the shores then untrod by the foot of any quadruped.

A characteristic of the tertiary formation of the Takaka Valley is large masses of fossiliferous limestone, beginning at the Tata Islands, and extending far up the valley. Under the limestone lies the Motupipi brown coal formation, which can be traced up the valley as far as Mr. Skeet's. I am indebted to Mr. James Burnett for carefully drawn and instructive plans and sections of the Motupipi working, which at once placed before me the character of the coal-field and the succession of the strata.* I need not here repeat what I have so

---

* SECTION OF STRATA AT MOTUPIPI, MASSACRE BAY.

No. 1 SHAFT.

		ft.	in.
Surface Clay			
Coal	... ...	1	9
Shale	... ...	2	0
ft. in.			
Coal ... 2 4 ⎫			
Band of Shale 0 5 ⎬ Working Seam	... ...	5	1
Coal ... 2 4 ⎭			
Shale of a sandy nature	... ...	3	5
Coal	... ...	1	3
Soft Sandstone, composed of very rough sand, like crushed quartz, with thin beds of shale; this stone falls away to loose sand under the pick, but stands very well in the shaft		7	0
Shale (pretty good roof)	... ...	2	0
Coal, with a great deal of water, sunk 2 ft. 7 in., and bored 2 ft. more to the bottom of this seam	... ...	4	7
		27	1
Bored 1 ft. further in shale	... ...	1	0
		28	1

Dip to the east, 1 in 17.

recently said at Auckland with regard to the quality and economical uses of this coal. The Motupipi coal is of the same geological age, and of the same description, as the Drury coal at Auckland. It is to be regretted that works commenced with so much judgment and regularity, and which might easily be continued, should be no longer carried on, in consequence of the high price of the fuel, and the difficulty of putting it on board the ship. To obviate the last difficulty, Mr. Burnet proposes a coal depot at the Tata Islands, where vessels could easily take it in.

Tertiary Formation of Blind Bay.

That the waters of Blind Bay at one time extended much further to the south, and covered a larger area, than they do now, is proved by the fact of a tertiary formation filling up the space between the eastern and western ranges from the lake country to the shores of the Waimea; but for the most part this formation is again covered by a more recent deposit of rolled stones, gravel, and diluvium, which at some places

SECTION OF STRATA AT MOTUPIPI, MASSACRE BAY.

No. 2 SHAFT.

	ft.	in.
Sea sand	4	6
Clay	1	0
Soft sandstone	10	8
Sandstone (very hard)	1	0
Shale (good roof)	4	7
Coal (hard and good, but at this place mixed with a little hard slaty shale, which decreases towards the south-west, and almost disappears at a distance of 7 ft. from the shaft)	4	4
Soft sandstone	5	0
Dark shale (almost black)	1	0
Soft sandstone	8	6
Sandstone (very hard)	0	6
Sandy shale	2	8
Dark shale	1	4
Slaty shale, mixed with coal	1	2
Bored through soft sandstone	3	6
	49	9

Full dip. about S. 75° E. 1 in 3.

Dip decreases with increased depth, and at bottom not more than 1 in 7.

This shaft was abandoned on account of the large quantity of water. Had this not taken place, it was intended to sink to a seam known to exist about low-water mark.

JAMES BURNETT.

attains a thickness of from 1,200 to 1,500 feet. So that the tertiary formation comes to the surface only just at the foot of the ranges, or where the rivers have cut through the diluvium and exposed the tertiary marls beneath.

I have before mentioned the fossiliferous marls, sandstones, and the coal-seams of the Wangapeka district; in the hills between the Buller River and the Rotorua Lake the same fossiliferous marls are met with. On the eastern side, near Nelson, the marine strata of the tertiary formation form the cliffs from Green Point to the Waimea Plains; and in a line between the town and the village of Richmond the brown-coal formation extends, opened up at Mr Jenkins' coal-mine.

The first excursion which I made after visiting Nelson was by the cliffs to this coal mine, and it was with great surprise that I saw the extraordinary disturbances which must have taken place in the stratification. The dip of the strata is towards the east, at an angle of about sixty degrees. As it is geologically impossible that a newer tertiary stratum can underlie the older slate formations of the ranges, it follows that the strata about Mr. Jenkins' coal-mine, by an immense force from the eastward, must have been completely turned over; and in the mine itself there is abundant proof of this. The strata there show unmistakable evidences that they have been rubbed and pressed together. Under these circumstances it is very doubtful whether a mine in that particular place could be successfully worked. To Mr. Jenkins the people of Nelson are indebted for proving to them that they have coal in the immediate vicinity of the town, and I think it not improbable that in the same line of stratification between Nelson and Richmond, a place may be found where, perhaps, at a greater depth, the coal seams lie in their natural position, and workings may be carried on with success.

The diluvial formation, which constitutes what is commonly known as the Moutere and Wai-iti hills, extending over an immense track of country towards the south, so far as to the Rotorua Lake, is nothing else than the detritus of the eastern and western ranges accumulated during a long period. It is not surprising, therefore, seeing that a portion comes from the auriferous rocks of the western ranges, that prospecting parties of diggers should have found gold at various spots between those hills. And it is a fact that the first gold in this Province was discovered in a stream which cuts through this diluvium. I refer to the Motueka diggings, in Pig Valley, at the foot of the western ranges. I have heard that quite recently these diggings had been resumed with some success. Bearing in mind the source from which the gold is derived, I think it likely that the nearer to the western ranges the richer will be the diluvium; but it is at the same time doubtful if it is rich enough, for any extent, to be of much consequence.

I have had many questions put to me with regard to the origin and character of the Nelson Boulderbank. I have not time to give such an explanation of it as I should wish to present to you. The boulders of which it consists are entirely syenite, and the same rock is found on the precipitous bluff which abuts upon the sea beyond Drumduan. The source is thus explained:—Fragments are constantly falling from the cliffs, and the action of the heavy northerly swell, combined with a strong current, takes them towards the south. The reason of their being deposited on the existing line is, that in all probability a submarine reef underlies them, of which the Arrow Rock, in the entrance of the Nelson Harbour, may be regarded as the southern termination. This supposition is strengthened by the fact of the Arrow Rock being of the same altered schists as occur immediately to the south of the syenite.

Before concluding, I wish to make a few remarks upon Volcanic Formations in the Province, and upon some general matters.

(5.) *Volcanic Formations.*

Although there are no signs of a volcanic action still going on in the Middle Island, as far as regards active volcanoes, solfataras, and fumaroles, like those in the Northern Island, we have at the same time plenty of proof that volcanic action has not been less powerful in the Middle Island than in the Northern.

I have not had an opportunity of visiting the volcanic districts of this island; but still, feeling that I should make some remarks upon this subject, I will endeavour to explain the opinion I have formed from specimens and communications.

It is well known that the high peaks of the Kaikoras, covered with perpetual snow, are of volcanic origin. My friend Haast describes the aspect of the three gigantic cones visible from the Awatere valley as most magnificent — three Mount Egmonts, one behind the other; the first one, Tapuaenuku (or Mount Odin), 9,700 feet high, a closed and rounded dome similar in shape to a cupola; the second one, further to the south, a truncated-bifurcated cone, the bifurcation undoubtedly the indication of a crater on the summit. From that peculiar form it has acquired the common name of the "Nest Mountain;" and the third gigantic peak, pyramidical in form like Mount Egmont. Almost equal in height to those landward Kaikoras is the lofty range which rises on the seaward side of the Clarence River, the principal points of which have been named Mount Thor and Mount Freya. Around these just mentioned Scandinavian monarchs of the mass, are ranged several smaller volcanic mountains, which I hope may have a right to maintain their Maori names.

It is not at all surprising that accounts have been received of newly-discovered hot springs in this volcanic region, in the Hanmer Plains,[*] at the foot of the Kaikoras.

Specimens forwarded to me by gentlemen, some from the Kaikoras, and some boulders from the Awatere, prove that the principal rocks in this district are basaltic and trachytic lavas.

Following the southerly direction indicated by the relative position of the two Kaikora groups, we come next to Banks' Peninsula, undoubtedly an extinct volcanic system, rising like an island out from the sea and level plains.

---

[*] At this house (Top-house), I met some gentlemen, newly arrived, and stopped on their road by the river. One had made the journey from Port Cooper, and he gave me an account of hot springs which he had discovered on the Hanmer Plains, under the shelter of a range of snow hills. He observed what seemed to him a remarkable fog, and, upon leaving his track to examine, he discovered some holes, which were filled with water of a temperature varying from milk-warm to almost boiling. The largest of them contained the hottest spring, and although he attempted to sound the depth with all the rope he could procure, he was unable to find the bottom. The circumference of the largest was about fifteen or eighteen yards. This is, I believe, the only instance of hot springs yet discovered in the Middle Island, and, if corroborated, may tend to throw some light on the volcanic connection between this district and the Northern Island.—"A Walk from the Wairau," from the *Nelson Examiner*, May 25, 1859.

In the same line, farther south, lies a third group of volcanic hills, forming the peninsula of Otago.

These three points doubtless indicate a line of volcanic action, running parallel to the great middle range or backbone of the Middle Island, on the eastern side of it. A closer examination, especially of the Otago Province, would no doubt furnish us with more and similar examples of volcanic hills on that line. It is remarkable, that—whilst the parallel zones of volcanic action on the Northern Island (the Tongariro zone, the Auckland and the Bay of Islands zone) all run on the western side of the backbone range between Wellington and the East Cape—the great volcanic line of the Middle Island takes the eastern side.

Many peculiarities in the physical features of the Middle Island, and also many interesting facts respecting the earthquakes in New Zealand, can be explained by the different position of the line of volcanic action in the two islands.

Reserving, however, for my future publications this interesting subject, which I regret I have not time to enter upon, I will, in conclusion, offer a few

GENERAL REMARKS.

In the earlier geological era of New Zealand, we may assume that both islands were connected, and that one backbone ran continuously from the South Cape to the East Cape. In the present map of New Zealand the integrity of this backbone is broken at Cook's Straits, and a closer inspection will show that there has been not only a simple break of continuity, but a lateral dislocation. Cook's Straits is, to use a miner's expression, a true fault. It is evident, from the rocks being of the same geological formation, that at one period the Pelorus ranges were a continuation of the Wellington ranges. The position of the strata in the eastern ranges of Nelson proves, that whilst the Northern Island seems to have remained stationary, some gigantic force has pressed the great mass of the Middle Island to the westward. The given description of the tertiary formation, extending between the ranges far up the valleys, sounds, and bays leading towards Cook's Straits on both sides, farther proves that the first act of this great conuvlsion of nature took place prior to the tertiary period, and the second and subsequent acts may be coeval with the period of volcanic action in the Islands.

While the tertiary sea was depositing the strata which now fill the valleys, and which rise in some parts to an altitude of 2,000 feet, the higher ranges of New Zealand only were above water.

Since the tertiary period, these islands have been gradually rising, and that rising has been coeval with the volcanic action, and developed to the greatest amount along the zones of volcanic action. It was in this time that the extensive plains on the east coast of the Middle Island, and the plateaus on the western side of the northern backbone, were raised above the sea.

The best proof of this rising of the land is to be found in the river terraces, which strike the eye of every traveller in the valleys of the Wairau, Awatere, Clarence, Motueka, Wangapeka, Buller, Takaka, and Aorere, and also in the lines of the sandy downs on the Port Cooper plains, which now, miles inland, mark the former limit of the sea.

These terraces are formed by the gradual rise of the land. If we suppose that, while the rivers are shaping out their beds, the upheaving movement is intermittent, so that long pauses occur, during which the stream will have time to encroach upon one of its banks, so as to clear away and flatten a large space, this operation being repeated at lower levels, there will be several successive cliffs and terraces. It is remarkable that in all the valleys the cliffs of the higher terraces are of greater altitude than the lower. At the Buller river, for example, near its outlet from the Rotoiti Lake, the uppermost cliff is a hundred feet in height, and there can be distinguished in one portion of the valley not less than eight terraces. The character of the terraces shows that the upheaving force has been decreasing towards the present time, either in power or period. The extreme height of these terraces, being not more than about two thousand feet up the valleys, shows the whole amount of rise in these islands, since the tertiary period, to be about two thousand feet.

Even at the present day, there are facts which prove that the land of these islands is not stationary, but that the relative levels of water and dry land are undergoing constant modifications. The rise of land at Wellington, in 1855, to an average height of three or four feet, over a great extent of coast, is familiar to every colonist. This rise of land, however, is not general over New Zealand; for there are many proofs that, while on the eastern side of the islands the level of the land is being raised, on the western side the land is sinking. An axis of equilibrium passes through the islands, on the western side of which the movement is downward, on the eastern side upwards. The same axis, curving round parallel to the Australian coastline, crosses the Pacific between New Caledonia and the Loyalty group, and can be traced through the Solomon Islands to New Guinea.

LADIES AND GENTLEMEN,—It now only remains for me to express my thanks to you for the attention with which you have followed my geological explanations. Much more still remains that I would wish to say, but I must now conclude.

I feel well assured that the mineral wealth of Nelson is not confined to what I have to-day mentioned, but believe that, in addition to gold, copper, and coal, future times will develop other valuable substances among your mountains and forests, which cannot fail to prove a source of wealth and prosperity to the Province of Nelson.

EXPLANATION OF THE MAP OF NELSON.
BY DR. F. V. HOCHSTETTER.
MAP VI.
THE PROVINCE OF NELSON, IN THE SOUTHERN ISLAND OF NEW ZEALAND.

AFTER a sojourn of seven months' duration on the Northern Island of New Zealand, I availed myself of the kind invitation of the Superintendent to visit the Province of Nelson, and devoted the months of August and September, 1859, to a geological survey of that Province. On the Southern Island I trod on a new and, compared with the Northern Island, an entirely different geological field, most remarkable on account of the multiferous mineral treasures, such as copper, gold, and coal, which have procured to the Province of Nelson the renown of being the principal mineral country of New Zealand. The fine and temperate climate of Nelson enabled me, even in the middle of winter, to pass and to explore the mountain chain which terminates near Cook's Straits. Into the higher and more distant regions of the Southern Alps, however, it was not possible to penetrate. From the Rotoiti Lake (Lake Arthur), the most southern part of which I visited, I saw the mighty summits of the southern mountain chains, covered with snow and ice, and which my friend and fellow-traveller Dr. Haast has since so successfully explored, with a most courageous perseverance, and under a great many difficulties and privations.* In the annexed map the results of his and my own observations are combined in a comprehensive delineation which explains the character of the geological structure of the northern part of the South Island.

From a centre which forms the division of the water-courses between the east and west coast, and which is the source of the two frontier rivers of the Provinces of Nelson and Canterbury, the Hurunui flowing east, and the Taramakau which runs to the west, there extend the two great mountain chains of the Southern Alps in a northerly direction through the Province of Nelson, terminating at Cook's Straits, where they give rise to the complicated coast-line which is so characteristic of the north extremity of the Middle Island.

Both of these mountain ranges differ in character. The western mountains, which end in Separation Point and Cape Farewell, have a direction from north to south. To them belong the Brunner chain, Lyell chain, Marino chain, Mount Owen, the Tasman mountains, and Mount Arthur chains; while to the north and fronting Golden Bay are the Whakamarama chain, Haupiri and the Anatoki chains. All these mountains and chains consist of crystalline and metamorphic rocks, of granite, gneiss, mica-schist, hornblende-schist, quarzite, and clay slate. It is to these rocks, which are auriferous, that Nelson is indebted for her gold-fields, which were the first gold-fields of New Zealand that were worked, and which even in 1859 yielded gold to the amount of £150,000

---

* J. Haast's Report of a Topographical and Geological Exploration of the Western Districts of the Province of Nelson. Nelson, 1861.

sterling. The nature of the gold-fields of the Aorere and Takaka valley convinced me that by a well-managed and regulated plan of working, and with a larger amount of capital, sufficient profits would be realised, and that the development of these gold-fields was only the commencement of gold discoveries which would ultimately extend throughout the whole mountain chain of the South Island; and that discoveries would be made which, though perhaps not equal to the gold-fields of California and Australia, will nevertheless class New Zealand amongst the gold lands of the earth.*

The summits of these mountain chains, such as the picturesque Mount Arthur, Mount Owen, and others, which are from 5,000 to 6,000 feet above the level of the sea, when covered with snow, are visible at a great distance. When arriving in Blind Bay, they give to the landscape of the Province of Nelson its peculiar charm. In the north of the Province are large plains, bordering important rivers, which intersect the mountain chains. Of these plains those by the Buller River are the most remarkable, where there is abundance of land fitted for agriculture, and rich natural pasture suitable for sheep runs. The western and south-western parts of the Province of Nelson are only now opened for settlement, and it may be surmised that in the next few years, these districts will become most important, on account of their treasures of coal, near the mouth of the principal rivers, viz., the Buller (Kawatiri) and the Grey (Mawhera).

The eastern mountain chain, running in a direction from south-west to north-east, consists of stratified sedimentary formations, comprising old grauwacke sandstone, red, green, and grey clay slates, and isolated patches of laminated calcareous strata. These strata, highly inclined, and trending in the same direction, are friction-breccias, accompanied by great masses of eruptive rocks, which have altered the contiguous strata. These eruptive rocks occur in a straight belt which extends from Stephens and D'Urville Island in Cook's Straits to the Cannibal Gorge in the south of the Province, over a distance of 150 miles. Throughout this line the lithological

---

* It will be remembered that the very rich gold-fields of the Province of Otago were discovered in 1861.

character of the rocks differs greatly according to the nature of the eruptive masses, which comprise varieties of Serpentine, Diabas, Syenite, Hypersthenite, and Augite Porphyry. To the belt of Serpentine and Hypersthenite belongs the celebrated Dun Mountain, the rich copper ores and chrome-iron from which have given rise to extensive mining operations.

These mountain ranges terminate at Cook's Straits in numerous islands and peninsulas, which inclose these fiord-like bays and sounds, Pelorus Sound and Queen Charlotte Sound, which even in Cook's time were celebrated as most excellent harbours. The mountains get gradually higher towards the south. Ben Nevis and Gordon's Knob, which are visible from Nelson, rise to an elevation of 4,000 feet. The mountain range is then broken for a short distance, rising however again in the immediate vicinity of the southern banks of the Rotoiti Lake to a much greater height, forming Mounts Travers and Mackay, and further, in a south-westerly direction, to a height of 10,000 feet in the Spencer Mountains (Mount Franklin and Mount Humboldt) much above the snowy line. This grand mountain chain forms a centre from which the principal rivers of the Province of Nelson have their source. It is to be regretted that the sandstones and clay-slates of this formation contain no fossils, and that the scanty traces of animal and vegetable remains which have been discovered give no certain clue to their geological age. There is only a single locality where fossils are to be found, and that is only at the extreme flank of the mountain system, near Richmond, a few miles south of Nelson, and these indicate a mezozoic age.

The country to the eastward of these mountain chains, from the Pelorus Sound to the Wairau Plains, and including the alluvial valleys of the Wairau, Awatere, and Waiautoa Rivers; also the great mountain ranges, 8,000 to 9,000 feet in height, of the seaward and landward Kaikoras, and the lofty peaks which have been named after the Scandinavian gods—Odin, 9,700 feet; Thor, 8,700; and Freya, 8,500—was withdrawn in 1859 from the Province of Nelson, to form the Province of Marlborough.

Between the eastern and western mountain ranges, is the deep indentation of the coast which forms Blind Bay, and from the southern extremity of which the land rises gradually to an altitude of 2,000 feet above the sea. Here are situated the picturesque mountain lakes Rotoiti and Rotoroa, at the points where the two ranges coalesce to form the Spencer Mountain, by which they are continued in a south-westerly direction. Near Nelson, the commencement of the highland is known as Moutere Hill, where it is intersected by innumerable ravines. This highland is composed of irregular layers of grit sand and yellow clay, resting on a tertiary formation containing brown coal, and filling up the contracting valley between the two mountain systems. These strata are of quaternary age, and being part of the generally diffused drift formation which fills up the principal valleys, and covers the flats amongst the mountains, afford evidence that only in the most recent geologic age they were covered by the sea. There is no doubt that the admirable climate of the shores of Blind Bay is due to the above described configuration. Even when there is a storm in Cook's Straits, it is calm and still in Blind Bay, being sheltered from the break of the sea by Separation Point and Cape D'Urville, while the strong southerly winds are broken by the mountain ranges which converge in that direction. Ships find in Blind Bay shelter from the dreaded storms that rage in Cook's Straits. The town of Nelson, situated at the southeastern border of the Bay, and at the base of the eastern mountain, enjoys, unlike the other ports of New Zealand, an agreeable absence of wind, and which, combined with a clear and rarely clouded sky, renders its climate the most agreeable and beautiful in New Zealand. With justice it may be called "The Garden of New Zealand."

The town of Nelson was established in 1842, and was the second settlement formed by the New Zealand Company in Cook's Straits. In spite of the grievous trials with which this young colony has had to contend, it has steadily gained ground. Thus, in 1843 it lost a great number of its best men in the bloody conflict with the natives when Rauparaha and Rangihaeata opposed the colonisation of the Wairau. However, through the exploration of the country, resulting in the discovery of coal, copper, chrome, graphite, and gold, Nelson has become the principal mineral-producing Province of New Zealand. Its population at this date amounts to 10,000 inhabitants, 5,000 of whom reside in the town and its vicinity. The town lies at the foot of the mountain, being built upon an alluvial delta formed by the confluence of two streams, named the Matai and the Brook Street Creek, extending also up their valleys and along the hill slopes that face the harbour. An excellent road leads from Nelson to the south through the luxuriant fields and meadows which bedeck the agricultural districts of the Waimea and Waiiti plains. On these fruitful alluvial flats are to be seen farm after farm, while many villages are rapidly springing into existence. Since 1861, Nelson has possessed a railway, being the first constructed in New Zealand. It is the work of the Dun Mountain Company, for the purpose of developing the chrome mines, and leads from the harbour through the town and up the Brook Street valley.

The existence of the harbour of Nelson is due to a most singular boulder bank which extends along the coast for eight miles, forming a natural dam, behind which there extends a narrow and shallow arm of the sea, which deepens towards the south, where it communicates with Blind Bay, and forms a small but safe harbour.

www.ingramcontent.com/pod-product-compliance
Lightning Source LLC
Chambersburg PA
CBHW030407170426
43202CB00010B/1518